LUNAR EXPLORATION
AND SPACECRAFT SYSTEMS

TARGET FOR LUNAR EXPLORATION
Surface of the Moon as Observed from Earth

AN AMERICAN **ASTRONAUTICAL** PUBLICATION
SOCIETY

LUNAR EXPLORATION AND SPACECRAFT SYSTEMS

Edited by
Ross Fleisig
Edward A. Hine
George J. Clark

Proceedings of the Symposium on
Lunar Flight
Held December 27, 1960, in New York City

Springer Science+Business Media, LLC

ISBN 978-1-4899-6215-7   ISBN 978-1-4899-6439-7 (eBook)
DOI 10.1007/978-1-4899-6439-7
© Springer Science+Business Media New York 1962
Originally published by American Astronautical Society, Inc. in 1962.
Softcover reprint of the hardcover 1st edition 1962

First Printing

Library of Congress Catalog Card Number: 62-13758

FOREWORD

This volume of the Special Astronautics Symposia presents the Proceedings of the American Astronautical Society's Lunar Flight Symposium. The symposium was held December 27, 1960, at the Biltmore Hotel in New York City as part of the 127th meeting of the American Association for the Advancement of Science.

The AAS New York Section sponsored the Lunar Flight Symposium as the AAS First Annual Eastern Regional Meeting. The symposium consisted of two technical sessions and a panel discussion and was under the over-all direction of Mr. Ross Fleisig (Program Chairman). Dr. Lisle L. Wheeler, Vice President for Engineering, Sperry Gyroscope Company, was Chairman of the Lunar Exploration Session and Mr. Robert Young, President of Budd Electronics, was Chairman of the Lunar Spacecraft Systems Session. Mr. Alfred M. Mayo, NASA Assistant Director for Bio-Engineering and 1961 AAS President, organized and moderated the Panel Discussion of the question: Is There a Need for a Manned Space Laboratory?

The Lunar Exploration Session was co-sponsored by the National Aeronautics and Space Administration. In this connection, the cooperation of Dr. Hugh L. Dryden, NASA Deputy Administrator, and Dr. Homer E. Newell, NASA Deputy Director, Office of Space Flight Programs, is gratefully acknowledged. Dr. Newell assisted in the review and selection of technical papers for this session.

The technical papers appear in this volume in the order of presentation at the meeting. They are presented basically in the form submitted by the authors, with only minor corrections being made by the editors. The recording of the panel discussion has been edited as well as reviewed by the moderator. In addition, each panel member has checked his contribution to the discussion.

FOREWORD

Committee members who made this meeting possible include: *Host Committee:* G.E. Pinkham (Chairman), AAS New York Section Chairman, Sperry Gyroscope Company; James H. Rosenquist, RCA Communications; *Preprint Committee:* Alphonse P. Mayernik (Chairman); Leroy McMorris; John J. Raffone, Bell Telephone Laboratories; *Publicity Committee:* H. E. Bockrath (Chairman), Grumman Aircraft Engineering Corporation.

Editor: Ross Fleisig*
Grumman Aircraft Engineering Corporation

Associate Editor: Edward A. Hine
Sperry Gyroscope Company

Associate Editor: George J. Clark
Grumman Aircraft Engineering Corporation

*Note: At the time of the meeting, Mr. Fleisig was affiliated with the Sperry Gyroscope Company.

CONTENTS

Scientific Objectives of Lunar Exploration
by Robert Jastrow.   1

Distribution of Dust in Cislunar Space — Possible
Existence of a Terrestrial Dust Shell
by Professor S. F. Singer.   11

Comparison of Special Perturbation Methods in Celestial
Mechanics with Application to Lunar Orbits
by Samuel Pines, Mary Payne, and Henry Wolf.   25

Radiation Shielding of Lunar Spacecraft by T. G. Barnes,
E. M. Finkelman, and A. L. Barazotti.   52

A Family of Radioisotope-Fueled Auxiliary Power Systems
for Lunar Exploration by Robert J. Wilson.   72

Extending the Range of Radar-Beacon Tracking for Lunar
Probes by Norman S. Greenberg.   94

Horizon Trackers for Lunar Guidance and Control Systems
by K. H. Kuhn and E. W. Stark.   108

Soft Lunar Landing Guidance Sensor by A. Barabush   153

Panel Discussion: "Is There a Need for a Manned Space
Laboratory?"   168

SCIENTIFIC OBJECTIVES OF LUNAR EXPLORATION*

ROBERT JASTROW

*Goddard Space Flight Center and Chairman, Lunar Science Subcommittee,
National Aeronautics and Space Administration, Washington, D.C.*

The NASA science program includes several major areas of research in which our efforts will be concentrated during the next ten years. They include studies of the earth, its atmosphere, disturbances on the sun and the effect they produce on the earth, investigations of the moon and the nearby planets, and studies of the stars and galaxies.

These projects draw on all the resources of chemistry, physics, astronomy, and the earth sciences. The union of these traditional disciplines into a new area of research marks a return to the old tradition of natural philosophy in which all the modern physical sciences had their roots. Scientists of many different backgrounds have been drawn together to work on these problems, united by a common interest in the structure of the world around them.

The vast increase in the number of papers on space research published in American journals in the last year bears witness to the development of a great interest in the new field of space research within the American scientific community. In one American scientific monthly, the *Journal of Geophysical Research*, the number of papers related to space research has grown from a half-dozen in 1958, to more than one hundred in 1960—a remarkable increase.

Among the varied activities of the space science program, the coming exploration of the moon and the nearby planets bears perhaps the greatest promise for providing fundamentally important and exciting discoveries in physics and biology, both for the scientist and for the layman. This particular set of projects also represents the true character of the NASA science program, more than does any other area of effort, because it is an *exploration science* and close to the spirit of the voyages of discovery and exploration in the 16th and 17th centuries.

The first extraterrestrial body to be explored will undoubtedly be the moon, which is the earth's nearest celestial neighbor. Mars and Venus are a hundred times more distant, and a rocket that could reach the moon in a day or two would take months to arrive in the

---

*This article is based in part on an article appearing in the May 1960 issue of *Scientific American*.

vicinity of one of these planets. An instrument station on the moon could also communicate with the earth more easily than one on Mars or Venus. The moon is a way station en route to the planets, and a testing ground for the development of rocket technology and scientific apparatus required for the exploration of the planets.

The moon is a body whose surface has probably preserved the record of its history remarkably well, and better in fact than either Mars or Venus. For this reason the moon is a particularly interesting object to the scientist, quite apart from its role in the technological development of the space program. It is a "Rosetta Stone" of the solar system, whose surface holds a key to the history and development of planetary bodies. The exploration of the moon will provide otherwise unobtainable clues to the manner of formation of the solar system, some 4½ billion years ago, and to its subsequent history.

It is a remarkable fact that all the craters on the moon, without exception, are precise circles or segments of circles. If on the earth we were to stake out a circle on the surface, and then return after a billion years, we would find that it had been distorted out of all resemblance to its original form by the displacements, upheavals, and other accompaniments of mountain-building activity which distort the earth's surface in periods of 10 to 50 million years. Not only the shape, but also the surface topography of any formation on the earth, would be destroyed by the erosion of wind and water, in approximately the same length of time. However, on the moon there is no water, no appreciable atmosphere, and in fact there is no way in which to produce substantial changes of any kind on the surface. It appears also, that there is little if any of the mountain-building activity which contributes to the destruction and replacement of large parts of the earth's surface, at a rapid rate. The circularity of the craters on the moon shows the absence of this mountain-building activity. For these reasons—the absence of erosion and the absence of mountain-building activity—the surface of the moon has preserved a record of its history which probably extends back through many billions of years to the earliest period of its existence and to the infancy of the solar system.

By virtue of the antiquity of its surface, the moon probably holds another remarkable record in the layer of cosmic and interplanetary dust which must have drifted down on it since the early days of its formation. We estimate that this dust may be as much as a foot or more in depth, and in it we may find organic molecules which could provide clues to the origin of physical life in the solar system. When we are able to reach the moon and carry out chemical analyses and radioactive dating of samples taken from the dust layer on its surface, we will find important clues to the chemical history of the solar system.

Presumably the record has been lost on Mars and Venus, which certainly have atmospheres, and which may have had oceans at an earlier stage in their histories. It is only on the moon that we· may have a reasonable expectation of finding this unique record of the history of the solar system preserved, and this is one of the primary reasons for the special importance of the moon in our space science program.

Program Objectives

The lunar science program for the next ten years divides into three areas of comparable importance.

First, *an extensive scientific exploration of the moon must be carried out*. Its surface must be mapped and analyzed according to the procedures used in the exploration of other virgin territories. Charts must be made of structural features which can affect navigability; the composition of the surface must be analyzed for minerals and other important constituents; and the data obtained from the exploration program must be integrated and applied to future projects for manned lunar exploration.

Second, *unmanned scientific observatories* must be set up to serve as monitors of solar weather and interplanetary properties over long periods of time.   These unmanned instrument stations should be designed for a useful life of several years or more.

Third, the *surface and interior of the moon* must be studied by apparatus based on the methods of geophysics in order to extract from this study those clues to the origin and history of major bodies in the solar system which we await with such interest. It should be noted that the primary emphasis in these studies will be on physical measurements, rather than on biological studies, because we are already certain that the moon does not provide a hospitable environment for living organisms, particularly because of the lack of water on and near its surface. There is in fact only the smallest probability of finding even the most primitive living forms on that body.

It is in the third area that we find the immediate scientific objectives of the lunar program. We have noted that the moon plays a special role in all investigations into the origin and history of the solar system. We are rather certain *when* the solar system came into being, namely, about 4½ billion years ago, but *how* it was formed is not known, and has been the subject of much thought and speculation for centuries. The exploration of the moon can make a unique contribution to the solution of this fascinating and important problem, and in the remainder of this discussion we will devote ourselves to aspects of the current plan for lunar exploration which have a special bearing on that problem.

Lunar Characteristics

Before entering on a discussion of specific scientific objectives, we first survey the current state of knowledge regarding the moon as a planetary body. The moon moves in an orbit which has the approximate shape of an ellipse around the earth and whose plane is inclined at 5 degrees to the orbital plane in which the earth moves around the sun, i.e., to the ecliptic. The mean radius of the moon's orbit is 232,000 miles; its period of revolution is 28 days; and its eccentricity is 0.055. The last figure means that the distance of the moon from the earth varies from 210,000 miles to 245,000 miles.

The radius of the moon is approximately 1000 miles, or one-quarter of the radius of the earth. Its mass is one-eightieth of the mass of the earth; the force of gravity at its surface is one-sixth of that on the surface of the earth. The mean density of the moon is 3.34, and this is the only bit of physical information regarding the bulk properties of the moon which is at present available. The moon's density happens to be close to that of another sample of extraterrestrial matter, a common type of stone meteorite known as chondrite, and this fact enters into current speculations which relate the composition of the moon to that of the meteorites.

The moon's shape is approximately spherical. Because of its rotation about its own axis once every 28 days, the moon should bulge out slightly at the equator under the influence of centrifugal forces. Calculations indicate that the amount of the bulge would be about 150 ft. Indirect measurements have shown that in reality the radius of the equator is approximately *1/2 mile* greater than the radius at the pole. Moreover, there is a bulge on the side of the moon that always faces the earth. This moon "nose" has been pulled out by the gravitational attraction of our planet. According to the calculations based on the known force of gravity, the nose should be about 120 feet high, but actually, as in the case of the equatorial bulge, it is much greater, namely, about 1/2 mile in height.

Thus the moon has a highly distorted figure in comparison with what our calculations would indicate, and this is a very significant result. The present distortions are those we would expect at an earlier stage in the history of the moon when it was closer to the earth, perhaps 50,000 miles away, and both the speed of rotation and the tidal forces of attraction to the earth were much greater. The enormous forces prevailing at that time must have molded the moon into its present shape. But these forces decreased as the moon spiraled out on a more distant orbit and turned more slowly on its axis. The fact that the moon has preserved the distorted figure of its infancy suggests that since its earliest days it has been at too low a temperature to provide the plasticity within its interior which would permit an adjustment to changing conditions. If the moon had been warm and plastic at any time during its history, the excess bulges described above would have sunk into its interior.

Measurements made with the help of artificial satellites have recently disclosed that the earth also departs significantly from its calculated shape. Bulges in the Northern and Southern Hemispheres give it a tendency toward the shape of a pear. Furthermore, the equatorial bulge is somewhat greater than was expected. In fact, the present equatorial bulge is what should have existed some 100 million years ago, when the earth was rotating faster and its day was shorter. Most scientists believe the meaning of these results is that the interior of the earth, unlike that of the moon, is sufficiently warm and plastic to respond to changes in rotation rate, but yet with enough resistance and internal strength so that its response lags behind the changes in rotation by an interval of about 100 million years.

Now let us consider some of the *surface characteristics* of the moon. These have a bearing on fundamental problems of lunar origin, and are also important for the preliminary design of lunar instrument stations. First, the *temperature* of the moon varies from a daytime maximum of 130° C at *noon*, that is, at full moon and directly under the sun, to a *nighttime* temperature of −150° C. Measurements on the variation of the temperature of the moon, at a single point on the surface during an eclipse, indicate that the change from the maximum day value to a value near the minimum night temperature takes place rather slowly, over the course of an hour or so. The slowness of the change indicates that the surface is a good insulator, and therefore is not compacted rock, but rather consists of loose dust or sand in a vacuum.

Emissions from the moon at radar wavelengths, which measure the subsurface temperature, indicate a relatively cozy value of -39° C at depths of a few inches to a foot below the actual surface.

The best lunar photographs are taken near the *terminator*, where the sun is low on the horizon and the long shadows bring out the relief in the surface topography. As a result, the impression has been generated of craggy peaks and ramparts rising precipitously from crater floors. The actual measurement of the angles of slopes, by the measurement of changes in the lengths of shadows when the sun rises over the lunar horizon, has shown that this general impression of the moon's surface is not accurate. The measurements show that rarely if ever on the moon's surface does the average slope exceed the relatively modest angle of 10 degrees. For example, Theophilus Crater, in the southwest section of the moon, is a substantial crater with a radius of approximately 65 miles; it is surrounded by circular ramparts rising to 15,000 ft; and there is a central massif whose peak reaches 1,500 ft above the crater floor. The Theophilus terrain presents a formidable appearance; however, the measured elevation profile, as revealed by a cross section through the center of the crater, demonstrates that the slopes are actually quite gentle, nowhere greater than 6 degrees on the average. On this basis, using all currently available information on the detailed fine structure of the moon's surface, we find that the floor of Theophilus Crater is a rolling landscape in which the formidable 15,000-ft ramparts can barely be discerned on the horizon. In reality the general terrain on the moon is even more monotonous, because the elevation profile was constructed on a flat base, without allowance for the curvature of the moon's surface—which would in fact lower the crater ramparts below the horizon.

Lunar Investigations

The scientific objectives of the lunar science program relate to the study of the moon as a planetary body. We must determine the bulk properties of the lunar interior, including the degree of differentiation of the matter within the moon, and the presence or absence of a dense core such as exists within the earth; the structure, composition, and physical parameters of the surface; the radioactivity contained in the surface; and the relation of all these properties to the temperature and stress histories of the moon. Our fundamental concern is with the bearing of these results on the origin and development of the moon as a planetary body, within the context of the evolving solar system.

The lunar science program begins with the delivery of simple instrument packages to the vicinity and surface of the moon, which will provide preliminary information on the surface structure, surface radioactivity, and level of seismic activity within the lunar interior. Later projects will deposit more complex instruments on the surface of the moon, for TV reconnaissance of the nearby terrain and for detailed chemical and physical tests of the lunar matter near the landing site. Developments still further in the future may involve the remote-controlled return of samples of lunar matter to the earth from the moon. We may also be able to deposit unmanned remote-controlled mobile vehicles, capable of sampling the composition of the surface over a large area around the landing site.

One series of experiments is of particular interest in the present discussion of scientific objectives, *first*, because it has a special bearing on the planetary properties of the moon and the information which we expect to derive regarding the origin and history of the solar system, and *second*, because it illustrates the important role which is played by the integration of many distinct experimental techniques in the attack on a single major scientific objective.

This is the question of the distribution of mass within the moon. We know that at the center of the earth there is a dense liquid core, probably a mixture of iron and silicates, about 2,000 miles in radius; surrounding this core is the mantle, which is also about 2,000 miles in radius and consists of solid rock under tremendous pressures. The mantle is three times less dense than the center core. Surrounding the mantle is a thin crust of rock and other material whose thickness ranges from 3 to 30 miles. There is a heavy concentration of radioactive elements in the crust.

This picture of the earth, which has been derived from the study of earthquake records and a great variety of other sources, shows that the earth is a highly *differentiated* body, both in chemical composition and in density. That is, the interior of the earth is not homogeneous.

Now we immediately ask whether there is a similar differentiation within the interiors of the moon, Mars, and Venus; and if there is some degree of differentiation within these bodies, how does it compare with that in the earth's interior; and finally, can we find out how these bodies were formed, and what their history has been, by comparing the degrees of differentiation within the interiors?

Our interest here is in whether the planets and their satellites accumulated out of the gas, dust, and debris of various sizes which must have existed in the cloud surrounding the primordial sun, or whether, perhaps, the planetary bodies were torn from the body of the sun during a near collision with a passing star—an alternative explanation of the origin of the solar system. We would like to know what the internal temperatures of the planetary bodies were at the time of their formation and in the early stages of their histories, and all other information related to their origin and development. The *temperature history* of a body such as the moon is particularly important. If the moon were formed by the condensation of a mass of hot gas ejected in a collision between the sun and another star, then it probably existed in a molten state at some early stage in its history. In this case the iron in its interior will have collected at the center in a dense core, as in the earth.

The first step is therefore to search for evidence of a dense core within the moon. We can attack this problem by placing a lunar satellite in close orbit around the moon. This satellite would contain many scientific instruments, but just the determination of the orbit of the lunar satellite would in itself provide a very important result, for if it could be tracked with high accuracy, the measured variations in its orbit would provide data on the gravitational field of the moon, just as the tracking of the Vanguard satellite has provided a measurement of the flattening of the earth at the poles, and has yielded the discovery of the pear-shaped component.

The combination of this lunar satellite information with previously existing astronomical information on the moon's motion about the earth will permit us to determine the mean moment of inertia of the moon, which means, in effect, the degree of concentration of mass at the center of the moon. Thus, if the moon has a dense core such as exists within the earth, we will find this out from the analysis of the lunar satellite orbit.

This experiment may tell us that the moon has, on the other hand, an *undifferentiated* structure in which bits of iron are distributed through the body of silicon and oxygen within the planet, like raisins in a fruitcake.

Of course, it is possible that the moon has a high concentration of the radioactive elements, uranium, thorium, and potassium, and that it could have been heated to the melting point at some time in the past by the energy released in the radioactive decay of these elements. If enough radioactive elements are present, we know iron will melt within its interior simply from the heat generated by these nuclear decays. In the case of the moon, we do not know the amount of radioactivity present. Therefore, one of our first and most important experiments in the lunar program will be the measurement of the radioactivity on the moon. A preliminary and very rough experiment of this kind is included in the earliest phase of the lunar project. Later phases will almost certainly include more sensitive measurements of lunar radioactivity. It is clear that these measurements of the level of radioactivity are essential to the determination of the moon's history.

Even if we can measure the mass distribution within the moon and the amount of radioactivity, there is still a further degree of complication in the interpretation of these results. The complication consists of the fact that radioactivity measurements only give the concentration of radioactive elements *on the surface*, and cannot give the *average* concentration within the interior. However, the range of possibilities regarding the distribution of radioactive elements within the interior will be narrowed if we measure the *flow of heat* through the moon's surface. This flow of heat will depend on whether the radioactive elements exist through the moon in the same concentration that they have on the surface. Clearly the generation of heat, and the flow of heat out through the surface, will be greater if the radioactive elements are distributed uniformly through the moon with the same concentration in which they occur on the surface, than if they exist in the surface layer alone.

Such heat-flow measurements are a part of the planning for the later stages of the lunar science program.

This particular set of experiments illustrates the manner in which one measurement in the lunar science program ties into another, and the combination of experiments is selected to bring about a vastly greater increase in our knowledge than could be realized from disconnected measurements.

Conclusion

The integration of the experiments, and their interpretation within the context of all other available information regarding the solar system, is essential for the achievement of this major scientific objective of the lunar program—the determination of the origin and history of the solar system. In formulating the integrated program, we have relied on scientists within the NASA, and in the universities, on Nobel laureates and members of the National Academy of Sciences, whose work is distinguished by a breadth of viewpoint and a deep understanding of all relevant fields of research. We will rely on these scientists in the future interpretation of the results to come out of the lunar program. With their continuing participation we hope to make rapid progress toward the solution to one of the most fascinating problems in modern science.

DISTRIBUTION OF DUST IN CISLUNAR SPACE— POSSIBLE EXISTENCE OF A TERRESTRIAL DUST SHELL*

PROFESSOR S. F. SINGER

Physics Department, University of Maryland, College Park, Maryland

The density of zodiacal dust in the plane of the ecliptic near the earth has the value of about 10^{-22} gm/cm^3. It is shown here that gravitational action of the earth can lead to substantial increases in densities in the vicinity of the earth. With the particles' orbits in the solar system similar to that of the earth (i.e., small geocentric velocities), enhancement factors of the order of 10-100 are possible; *a peak* in the concentration is reached at an altitude of about 1,000 km above sea level. The rate of accretion of interplanetary dust into the earth's atmosphere is also increased by this gravitational action.

For small dust particles ("smoke") additional forces become important and determine their orbits. These are the forces of radiation pressure, magnetic forces and, above all, the electric drag of a charged dust particle moving through the ionized outer atmosphere of the earth. A peculiar effect occurs which leads to a reversal of the electric charge of the dust particle in the vicinity of two earth radii, or about 4,000 miles above sea level.

Because of this effect the concentration of dust particles would begin to be very pronounced at about 4,000 miles altitude, and increase toward the earth with a maximum at some distance above sea level. This phenomenon may be identified with a "dust shell".

An interesting effect is noted, namely, that at about 4,000 miles the dust particle density may actually exceed the density of the atmosphere, both measured in gm/cm^3. Because of this effect the dust particle belt may have a profound influence on the radiation belt consisting of high-energy protons. It is possible that the dust belt limits the intensity of the proton radiation belt; in fact, the decrease of proton intensity between 1½ and 2 earth radii may be ascribed to the "sweeping" action of the dust particle belt, although other explanations cannot be excluded and may in fact be more likely.

The existence of interplanetary dust of dimension of 10-100 microns has been determined from observations of the zodiacal light

---

*This work was supported by a grant of the National Aeronautics and Space Administration.

by van de Hulst [1] and by Allen [2]. Opik [3] has given an analysis of the dynamics of the interplanetary dust cloud and its dissipation. Whipple [4] and the present author [5] have both discussed the problems which the dust environment raises for space flight through the erosion of the surface of satellites and space vehicles.[1]

The dust particles, sometimes also called *micrometeorites*, are thought to consist of small grains (either stony or metallic). They are not thought to possess the structure of meteors, which are most likely a conglomeration of such grains in the form of dust balls, as discussed by Whipple [6]. The micrometeors are too small to produce any visible effects and hence are not detected from the ground as they enter the earth's atmosphere.

Therefore, their distribution in the vicinity of the earth must be deduced either by direct observation through measurement of particle impacts on sensitive instruments in space probes, or by theoretical methods. The present paper attempts to provide a theoretical basis for calculating the dust particle density near the earth.

Forces Acting on Dust Particles

The distribution of dust particles must be determined by the forces which act on them. In cislunar space the gravitational force of the earth predominates and is at least two orders of magnitude greater than any other force for the particles concerned. However, there are three other forces which need to be taken into account: (1) radiation pressure, (2) the Lorentz force produced by the earth's magnetic field, and (3) a coulomb force produced by the interaction with the ionized exosphere.

The radiation pressure force is well known and is due chiefly to the solar radiation and, to a smaller extent, to the earth's radiation. The Lorentz force acting on a charged dust particle is analogous to the forces acting on cosmic rays in the earth's magnetic field. The possibility of this geomagnetic control on dust particles seems to have been first recognized by Dubin [7] and was used in his attempts to account for the formation of the sporadic E-layer in the ionosphere through the incidence of interplanetary dust.

The coulomb force or electrostatic drag force was discussed by the author [8] and shown to be one of the most important forces for

---

[1]Whipple [21] has also pointed to the psychological hazards on human inhabitants of space vehicles which might be produced by the noise of dust particle impacts.

particles moving in interplanetary space. A similar conclusion was reached by Opik [3]. This electrostatic drag force is important for any charged body moving through a plasma and is therefore also effective for earth satellites, but not nearly as effective as it is for small bodies of the size of dust particles [9-13].

The importance of the coulomb force hinges on the question of the charge of the dust particle. This possibility of charged dust was first pointed out by Whipple, in 1942, and by Spitzer. The dynamics of the situation were investigated by the present author [8], who also discussed the competition between photoelectric emission and accretion of electrons from the surrounding plasma. It was concluded that in interplanetary space the dust particles would most likely be positively charged. Whipple [14] has explicitly pointed out that, if one adopts the point of view of competition between photoelectric effect and ambipolar diffusion, then the charge becomes less positive as one approaches the earth and becomes negative in the ionosphere.

We propose here to make use of the following effect, namely, that a dust particle can have its orbit perturbed by the coulomb drag so that it ends up in a nearly circular orbit in a region where its charge is close to zero, at about two earth radii.

In this region it can then survive for a long period of time and therefore contribute appreciably to the density of dust, thereby forming a dust belt. This possibility is discussed with the help of a simple mathematical model. No great physical reality can as yet be claimed for the model, but we believe that effects of this general type are acting in cislunar space and can lead to a large concentration of dust at a distance of some thousands of kilometers above sea level.

Gravitational Concentration of Dust

We shall first neglect all forces except gravitational ones; hence, the results given here apply to large dust particles, certainly for those of the order of 100 microns. It is assumed that the geocentric velocities of these dust particles are isotropically distributed. This means that at large distances from the earth, where the gravitational potential due to that body is negligible, the particles have randomly oriented velocities with respect to it. Furthermore, these velocities are small, of the order of 5 km/sec. This conclusion is reached by considering the effects of the Poynting-Robertson effect on the orbits of small dust particles [15, 3].

Our treatment of the gravitational effect [16] leads to an enhancement of dust density in the vicinity of the earth, with a peak in density at some distance above sea level. The analysis has been carried out using the principle of Liouville's theorem from statistical mechanics. The result in terms of the density at infinity is given by the following expression:

$$N(y)/N_\infty = \frac{u^2 + y}{2u^2} \left[1 + \left(1 - y^2\, \frac{u^2 + 1}{u^2 + y} \right)^{1/2} \right] \qquad (1)^2$$

where $y = R/r$, where r is the distance from the center of the earth and R the radius of the earth. Here u is the geocentric velocity of the dust particle (at large distances from the earth) measured in terms of the escape velocity from near sea level, i.e., 11 km/sec.

The function is plotted in Figure 1 and shows some extremely interesting features. In the first place it is seen that the density in cislunar space, in particular close to the earth, is always enhanced over the density in interplanetary space. The enhancement is particularly pronounced for small values of u. The remarkable feature, however, is the fact that a pronounced peak always exists above the atmosphere at about an altitude of one-tenth of an earth radius for $u = 0.1$, and slowly moves to higher altitudes for larger values of u. When u exceeds 1.0 the intensity very close to the earth drops *below* the free-space density.

---

[2]Note added in proof:

Equation (1) should read

$$N(y)/N_\infty = \left(\frac{u^2 + y}{4u^2} \right)^{1/2} \left[1 + \left(1 - y^2\, \frac{u^2 + 1}{u^2 + y} \right)^{1/2} \right] \qquad (1)$$

This change modifies some of the subsequent equations as well as Fig. 1 (in detail), but does not alter any of the conclusions reached in this section. Note also that we refer here to dust *concentration* and not to the *flux* recorded by an impact detector. For a fuller account of this subject, see Ref. [16], as well as a note on Interplanetary Dust Near the Earth, *Nature*, 192 (1961) p. 321. In these publications we refer, more properly, to a dust *shell*, rather than a *belt*; we adopt the terminology of "smoke particle" for bodies small enough to be affected by radiation pressure and electromagnetic forces.

This can be seen more clearly by setting $y = 1$ in Eq. (1). We then have

$$N(1)/N_\infty = (u^2 + 1)/2u^2 \tag{2}$$

which for small values of u reduces to $(2u^2)^{-1}$ and for large values of u reduces to $1/2$.

For small values of u we can replace Eq. (1) by

$$N(y)/N_\infty = (2u^2)^{-1} y [1 + (1 - y)^{1/2}] = N_R y [1 + (1 - y)^{1/2}] \tag{3}$$

where N_R is the density very close to the surface of the earth, i.e., just above the appreciable atmosphere. We can investigate the re-

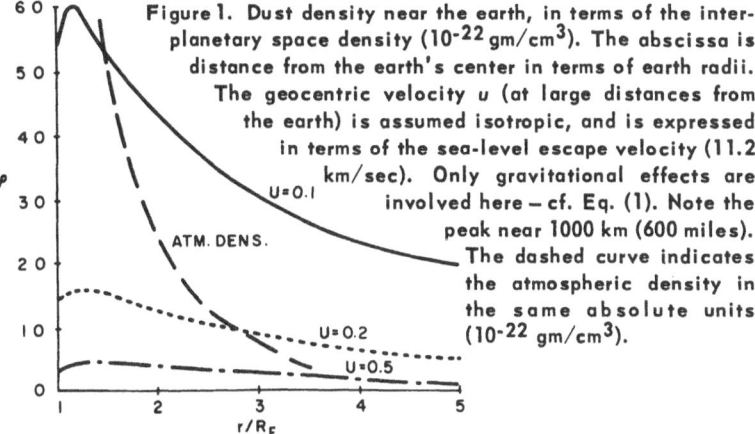

Figure 1. Dust density near the earth, in terms of the interplanetary space density (10^{-22} gm/cm^3). The abscissa is distance from the earth's center in terms of earth radii. The geocentric velocity u (at large distances from the earth) is assumed isotropic, and is expressed in terms of the sea-level escape velocity (11.2 km/sec). Only gravitational effects are involved here – cf. Eq. (1). Note the peak near 1000 km (600 miles). The dashed curve indicates the atmospheric density in the same absolute units (10^{-22} gm/cm^3).

mainder of expression (3). It reduces to 1 at $y = 1$. If we replace y by $1 - \epsilon$, then the expression reduces to $1 + \epsilon^{1/2}$. Therefore we see that there is always a peak below $y = 1$, and hence at some altitude above sea level.

We also investigate the rate at which dust particles are accreted by the earth, this rate being given by

$$N_\infty (\pi R^2) u (1 + v_\infty^2/u^2) \tag{4}$$

The expression is seen to have a minimum value when the geocentric velocity u equals the sea-level escape velocity v_∞. From the equa-

tion of continuity we can also deduce the mean radial velocity near
sea level. It is equal to Eq. (4) divided by N_R

$$V_{rad} = \frac{1}{2} \frac{u(u^2 + v_\infty{}^2)}{u^2 + 1} \tag{5}$$

Analysis of Perturbing Forces

In addition to the gravitational force the following perturbing
forces act on the dust particle. These have been discussed earlier
[8], but now will be applied to the case near the earth and discussed
in detail.

Coulomb Drag. We will follow a treatment by Spitzer [17]. The
drag deceleration is given by

$$a_c = - A_D \left(\frac{m_1}{2kT_1}\right)\left(1 + \frac{m}{m_1}\right)G(l_1 w) \tag{6}$$

where

$$A_D = 8\pi e^4 n_1 Z_1{}^2 Z^2 \ln \Lambda/m^2$$

$$G(x) = \frac{\Phi(x) - x.\Phi'(x)}{2x^2}$$

$$\phi(x) = \frac{2}{\pi^{1/2}} \int_0^x e^{-y^2} dy$$

$$\Lambda = \frac{3}{2ZZ_1 e^3}\left(\frac{k^3 T^3}{\pi n_e}\right)^{1/2}$$

$$l^2 = m/2KT$$

$$x = l_1 w = (1.5)^{1.5} w/C_1$$

where C_1 is the rms velocity $(3kT_1/m_1)^{1/2}$ and subscript 1 refers
always to the field particles. For our case

$$T_1 = 1500^\circ K; C_1 = 6.1 \text{ km/sec for protons}$$
$$= 260 \text{ km/sec for electrons} \tag{7}$$

We take $Z_1 = 1$; $\ln \Lambda \sim 20$; $G(l_1 w) \sim 0.2$ for the field protons, and less than 0.015 for field electrons. Since the dust particle mass $m \gg m_1$, Eq. (6) reduces to

$$a_c = 1.23 \times 10^{-23} Z^2 n_1 / m \tag{8}$$

The drag force F_D is given by $1.23 \times 10^{-23} Z^2 n_1$ dynes. The particle has mass m, radius a, density ρ. Its charge $Z = 6.95 \times 10^6 aV$, where V is measured in volts, and is given as a function of r/R_E from an analysis involving competition between photoelectric effect and ambipolar diffusion [8]:

$$V = 7.5 \, (r/R_E) - 15 \text{ (in volts)} \tag{9}$$

The ambient electron density above about 1.5 R_E is given by Helliwell:

$$n_1 \cong 10^4 \, (r/R_E)^{-3} \tag{10}$$

so that

$$F_D = 1.23 \times 10^{-23} \, [(6.95 \times 10^6)^2 a^2 (7.5 \, r - 15)^2] \, (10^4 \, r^{-3})$$

$$\sim 3.4 \times 10^{-4} \, a^2 \left(\frac{1}{r} - \frac{4}{r^2} + \frac{4}{r^3} \right) = k a^2 \, r^{-1} f(r) \tag{11}$$

where r is expressed in terms of R_E and $f(r) = (1 - 4/r + 4/r^2)$. Near sea level $f(r) = 1$; at $2R_E$, $f(r) = 0$.

The *Coulomb deceleration* can be expressed as

$$a_c = 8.1 \times 10^{-5} \, r^{-1} f(r) \, (a\rho)^{-1} \text{ cm/sec}^2 \tag{12}$$

This acceleration may be compared with the accelerations due to *Radiation pressure*:

$$a_{RP} = 1.5 \frac{1400 \text{ w/m}^2}{c} \frac{3\pi a^2}{4\pi a^3 \rho} = 3.5 \times 10^{-5} \, (a\rho)^{-1} \tag{13}$$

Gravity:

$$a_G = 980 r^{-2} \tag{14}$$

Magnetic:

$$a_M = qwB/m \tag{15}$$

Using mKs units, $q = 1.6 \times 10^{-19} Z$, $w \sim$ orbital velocity $\sim 10^4 r^{-1/2}$, $B \sim 5 \times 10^{-5} r^{-3}$ webers/m^2

$$a_M = 8 \times 10^{-15} (Z/m) r^{-3.5}$$
$$\sim (a^2 \rho)^{-1} (r - 2) r^{-3.5} \times 10^{-7} \text{ cm/sec}^2 \tag{16}$$

Integration of Dust Motion (using Coulomb Drag)

The energy loss per orbit in a circular orbit is $\Delta E = 2\pi r F_D$ and this will reflect itself in a lowering of the orbit by an amount Δr in a time Δt which is simply the orbital period:

$$\Delta t = T = 2\pi (GM_E)^{-1/2} r^{3/2} \tag{17}$$

Since the total (kinetic plus potential) energy is given by

$$E = \frac{1}{2} \frac{GM_E m}{r}$$

so that
$$\frac{dE}{dr} = -\frac{1}{2} \frac{GMm}{r^2}$$

$$\Delta r = (-\Delta E) \frac{2r^2}{GMm} \tag{18}$$

We can therefore obtain an expression for the rate of shrinkage of a circular orbit, as follows:

$$\Delta r / \Delta t = 2\pi r F_D \times \frac{2r^2}{GMm} \frac{(GM)^{1/2}}{2\pi r^{3/2}} = \frac{2}{(GM)^{1/2}} \frac{F_D}{m} \times r^{3/2} \tag{19}$$

$$= \frac{1}{\pi} T \frac{F_D}{m}$$

which is $(1/\pi)$ times the orbital period times drag deceleration.

The contribution of a single dust particle to the dust particle concentration can now be evaluated as follows. It is proportional to $\Delta t / \Delta r$ and inversely proportional to the volume $4\pi r^2 \Delta r$. Therefore

$$N(r) = C \frac{(GM)^{1/2}}{2r^{3/2}} \frac{1}{4\pi r^2} \frac{m}{F_D} = C \frac{(GM)^{1/2} m}{8\pi a^2 d} \frac{1}{r^{5/2}} \frac{1}{f(r)} \tag{20}$$

We can now examine the radial distribution of dust concentration due to particles which are captured into circular orbits with initial perigees beyond, say, $3R_E$. This includes particles[3] which have an impact parameter $b > 3R_E$ and are sufficiently small, i.e., $a < a_c$ as determined from Eqs. (12) and (14). We investigate the function

$$[r^{5/2}f(r)]^{-1} = \left[r^{5/2}\left(1 - \frac{4}{r} + \frac{4}{r^2}\right)\right]^{-1}.$$ It goes to infinity at the value

of r_0 where $Z = 0$. For the example considered, this occurs at $r_0 \sim 2R_E$. This infinity cannot, of course, be physically realized. There are various factors in existence which will keep this from happening. Among them are (1) effects of the neutral atmosphere; (2) magnetic field; (3) radiation pressure; (4) nonequilibrium perturbation of the particle's electric charge: (a) variations in the ionized exosphere; (b) variations in solar ultraviolet output (including day-night effect due to the earth's shadow); (c) charging effects by radiation belt electrons.

Special Cases

A discussion of all these factors is too difficult, and depends in many cases on the orientation of the particle's orbit. Let us therefore discuss a near-equatorial orbit of about $r_0 \sim 2R_E$. It can be seen by examining the magnetic force that only a clockwise orbit (as viewed from the North Pole) is stable. With $F_M = Ze\,(v \times B)$, a positive particle beyond r_0 will be deflected inward; then, as it passes within r_0 it becomes negative and is deflected outward.

The neutral drag force is $F_D \sim a^2\,(nm_H)v^2$, which should be compared with Eq. (11). It is seen to be of smaller orders of magnitude (for $a \lesssim 10\text{-}100$ microns). (The neutral gas density is of the order of the ion density.)

By far the most important effect appears to be the day-night effect. When the particle is in the earth's shadow its potential becomes negative. At $2\,R_E$ in an equatorial orbit the particle spends 1/6 of its time in the umbra. Its potential will be close to the value it has in the lower ionosphere, but because of the reduced value of the ion density the drag is reduced by $(r/R)^{-3}$. The aver-

---

[3]Particles with $a < a_c$ and $b < R_E$ do not contribute appreciably, while with $R_E < b < 3R_E$ the contribution is calculated separately.

age drag force over the orbit is therefore 1/6 x 1/8 = 1/48 of the drag in the ionosphere. With

$$\frac{1}{r} f(r) = \frac{1}{48} \qquad f(r) = 0.04 \qquad r^{5/2} f(r)^{-1} = 4.4$$

and the maximum concentration is reduced to 4.4 times the near-earth value, as shown in Figure 2.

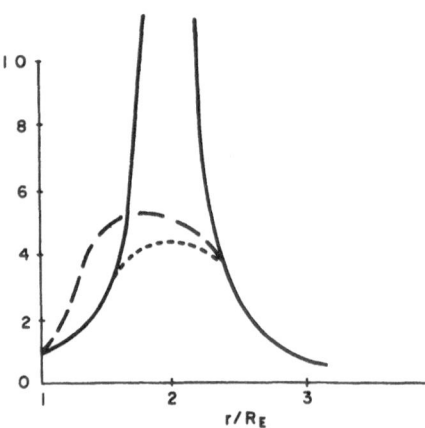

Figure 2. Sketch of the function $[r^{5/2} f(r)]^{-1}$ − cf. Eq. (20) − representing the contribution to the density of a single dust particle spiraling inward in a nearly circular orbit under the influence of Coulomb (electrostatic) drag. The dotted curve shows the influence of the day-night effect on the charge of the dust particle (assumed in an equatorial orbit). The dashed curve indicates qualitatively the contribution of particles captured into orbits with perigee less than $3R_E$.

Again it must be emphasized that what is calculated here is the contribution of a single particle which is captured into a nearly circular orbit above $3R_E$. Particles which are injected directly below $3R_E$ will add directly to the density. Their contribution is being evaluated by studying their detailed orbits.

Dust Particle Density as Deduced from the Radiation Belt

As can be seen from the foregoing treatment, detailed numerical integrations of orbits may be required in order to establish the exact distribution of dust particle density in the vicinity of the earth. This procedure is now being followed, and we are calculating a series of orbits, taking into account the appropriate forces as described previously.

A different approach, however, is also possible and it is this approach that we would like to expound now. Work on the distribution of trapped protons in the inner radiation belt has established that the source of the protons is most likely to be cosmic ray albedo neutrons [18]. The equilibrium intensity of the protons is controlled to a large measure by the density of the ambient atmosphere. According to this view, if the atmospheric density were the *only* pertinent factor, the proton intensity should increase continuously with altitude. It was surmised quite early, however, that the proton intensity should show a broad maximum between 1 and 2 earth radii [19]; this prediction was based on the idea that the trapping lifetime would be set by the limited ability of the magnetic field to trap particles of large radius of curvature, i.e., high energy protons. The existence of such a maximum has been established experimentally. While this may support the theoretical ideas with regard to nonadiabatic trapping of high-energy protons, another explanation is possible, at least a priori.

We could assume that the concentration of dust particles is sufficient to determine the intensity of the trapped radiation-belt protons in the vicinity of 2 earth radii. That this idea is indeed plausible can be seen from the gravitational concentration factors which were obtained earlier. For example, for $u = 0.1$ the gravitational concentration at 2 earth radii is 43.5, corresponding to a mass density of at least 4.35×10^{-21} gm/cm^3, while the atmospheric density at the same location is very likely to be only half that. Therefore, even gravitational concentration could give a density greater than atmospheric beyond about 1.6 earth radii. Now it is probably not reasonable to assume $u \lesssim 0.1$, and therefore gravitational concentration by itself may not be sufficient to determine the density in this region. However, our previous argument shows that if electromagnetic forces are taken into account, in addition to gravitational forces, the effects will certainly be enhanced.

Instead, we will turn the process around and use the fact that the trapped proton intensity is depressed in the vicinity of 2 earth radii, and to argue that the dust particle density there is perhaps an order of magnitude greater than the atmospheric density, using it to establish a normalization for the curve of Figure 2. This density will then be of the order of 2.6×10^{-20} gm/cm^3. This enhancement by a factor of about 200-1,000 over the interplanetary value is consistent with the value of 100-2,000 which can be determined from available satellite data on meteor dust particle impacts [21].

While a firm conclusion is as yet premature, it does appear that the dust particle density may, at certain locations, reach a value greater than the atmospheric density, and therefore control the intensity of the radiation belt, essentially by "sweeping out" particles which would otherwise remain trapped. Further work is urgently needed, both as to the theory of dust particle trapping, and, experimentally, on dust particle impacts. It can also be seen that experimental measurements of the energy spectrum of radiation-belt protons can be very helpful in deciding what physical mechanism actually causes the decrease of intensity of the protons beyond about 1½ earth radii. If it is really caused by a large dust particle density, then this will affect protons of all energies equally, so that the energy spectrum will not change with altitude.

Discussion of Gravitational Concentration

Dr. Fred Whipple has kindly called my attention to a paper by D. B. Beard [20] which deals with the same problem, namely, the gravitational concentration of dust particles in the vicinity of a planet. Beard's results, however, are quite different from our calculation [Eq. (1)]. He shows an enhancement which is about 1,000 times greater than ours near the earth, but without a peak. We believe this disagreement can be resolved as follows.

In Beard's paper it is assumed that the particles have initial geocentric velocities (and should therefore execute hyperbolic orbits near the earth); but these orbits are modified by the presence of a third body, the sun. This would correspond to so-called "Jacobi capture"; however, it is well known that only a small fraction of the particles—namely, those entering with the right velocity and the right angle with respect to the sun-earth line—can be put into bound orbits and captures. However, this problem is not treated; instead, the author simply assumes that the particles everywhere have a radial kinetic energy equal to the potential energy—an assumption which cannot be supported by physical arguments. In addition, the tangential velocity is neglected in the subsequent treatment, i.e., it is implicitly assumed that the angular momentum of every particle is zero. Under this highly artificial assumption, all particles would naturally fall in toward the earth along radial orbits with parabolic velocity. The variation in density is then given very simply by the area decrease, which goes as r^{-2}, divided by the velocity which goes as $r^{1/2}$. Therefore the concentration of particles varies as $r^{-3/2}$. If the particles started from infinity, the density would be infinite near the earth, as would the rate of accretion; however, the particles are supposed to be introduced at a distance of 100 earth radii, in which case the density near sea level will be $100^{3/2}$ or 1,000 times the space density. Clearly, had the particles been introduced at a distance of 10 earth radii, the density would only have been 31 times the space density.[4]

In a recent paper Whipple [21] considered the problem of how the dust particle density might be increased near the earth. He rejects mechanisms which produce the particles from the earth itself and proposes a novel mechanism which would increase the number of interplanetary particles near the earth without increasing the mass density. This mechanism is based on the breakup of zodiacal particles into smaller ones through electrostatic forces. While this mechanism may be applicable to the larger meteors which consist of weakly bound grains, the small dust grains themselves may be quite strong and require high electric charges.

Whipple [21] has also recognized the importance of electrostatic drag as a means of enhancing the density near the earth. This, in fact, is the main mechanism which is investigated in our present paper.

References

1. H. C. van de Hulst, "Zodiacal Light in the Solar Corona," *Astrophys. Journ.*, 105 (1947), pp. 471-488.

2. C. W. Allen, "The Spectrum of the Corona at the Eclipse of 1940 October 1." *Monthly Notices Roy. Astron. Soc.*, 106 (1946), pp. 137-150.

3. E. J. Opik, "Interplanetary Dust and Terrestrial Accretion of Meteoric Matter," *Irish Astron. Journ.*, 4 (1956), pp. 84-135.

4. F. L. Whipple, "The Meteoritic Risk to Space Vehicles," in *Proceedings of the Eighth International Astronautical Congress* (Barcelona, 1957), Vienna, Springer-Verlag, 1958, pp. 418-428.

5. S. F. Singer, "Effects of Interplanetary Dust and Radiation Environment on Space Vehicles," in *Physics and Medicine of the Atmosphere in Space*, ed. O. O. Benson, Jr., and H. Strughold, New York, Wiley, 1960, pp. 60-900.

---

[4]The rate of accretion according to Beard [20] would be $N_\infty \left(\dfrac{r_0}{R}\right)^{3/2} v_\infty$
$\times 4\pi R^2$ in our notation. It is seen to depend on the choice of r_0. For $r_0 = 100R$ the accretion $(10^{-22} \text{ gm/cm}^3) (100)^{3/2} (11.2 \times 105 \text{ cm/sec}) 4\pi$ $(6.37 \times 10^8 \text{cm})^2 = 5.7 \times 10^5 \text{gm/sec} = 48,000 \text{ tons/day}$.

6. F. L. Whipple, and G. S. Hawkins, "Meteors," in *Handbuch der Physik*, ed. S. Flugge, Berlin, Springer-Verlag, 1959, **52**, pp. 519-564.

7. M. Dubin, "Meteor Ionization in the E-Region," in *Meteors*, ed. T. R. Kaiser, London, Pergamon Press, 1955, pp. 111-118.

8. S. F. Singer, "Measurements of Interplanetary Dust," in *Scientific Uses of Earth Satellites*, ed. J.A. Van Allen, Ann Arbor, Mich., University of Michigan Press, 1956, pp. 301-316.

9. B. Lehnert, "Electrodynamic Effects Connected with the Motion of a Satellite of the Earth," *Tellus*, **8** (1956), p. 408.

10. R. Jastrow and C. A. Pearse, "Atmospheric Drag on the Satellite," *Journ. Geophys. Res.*, **62**, (1957), pp. 413-423.

11. K. P. Chopra and S. F. Singer, "Drag of a Sphere Moving in a Conducting Fluid in the Presence of Magnetic Field," in *1958 Heat Transfer and Fluid Mechanics Institute*, Palo Alto, Stanford University Press, 1958, pp. 166-175.

12. L. Krauss and K. M. Watson, "Plasma Motions Induced by Satellites in the Ionosphere," *Phys. Fluids*, **1** (1958), p. 480.

13. D. B. Beard and F. S. Johnson, "Charge and Magnetic Field Interaction with Satellites," *Journ. Geophys. Res.*, **65** (1960), pp. 1-7.

14. F. L. Whipple, "Meteoritic Material in Space," in *Physics and Medicine of the Atmosphere and Space*, ed. O. O. Benson, Jr., and H. Strughold, New York, John Wiley and Sons, 1960, pp. 48-59.

15. F. L. Whipple, "A Comet Model; III, The Zodiacal Light," *Astrophys. Journ.*, **121** (1955), pp. 750-770.

16. S. F. Singer, "Gravitational Concentration of Interplanetary Dust," *Journ. Geophys. Res.*, 1961.

17. L. Spitzer, Jr., *Physics of Fully Ionized Gases*, New York, Interscience, 1956, pp. 73-81.

18. S. F. Singer, "On the Nature and Origin of the Earth's Radiation Belts," in *Space Research*, ed. H. K. Kallmann-Bijl, Amsterdam, North Holland Publ. Co., 1960.

19. S. F. Singer, "Trapped Albedo Theory of the Radiation Belt," *Phys. Review Letters*, **1** (1958), pp. 181-183.

20. D. B. Beard, "Interplanetary Dust Distribution," *Astrophys. Journ.*, **129** (1959), pp. 496-506.

21. F. L. Whipple, "Particulate Contents of Space," in *Medical and Biological Aspects of the Energies of Space*, ed. P. Campbell, New York, Columbia University Press, 1961.

COMPARISON OF SPECIAL PERTURBATION METHODS IN CELESTIAL MECHANICS WITH APPLICATION TO LUNAR ORBITS

SAMUEL PINES, MARY PAYNE, AND HENRY WOLF*

*Applied Mathematics Department, Research and Development Division,
Republic Aviation Corporation, Farmingdale, New York*

The object of this investigation is to make a critical comparison and evaluation of three commonly used methods of special perturbations. The actual orbits used for this comparison were typical of lunar shots. Although conclusions apply directly to lunar orbits, more general conclusions may also be drawn.

The methods considered are those of Cowell and Encke, as well as the device of the variation of parameters. Also included is an evaluation of two numerical integration schemes, a fourth-order Runge-Kutta-Gill, and a sixth-order backward difference scheme.

Rather than compare the results of these methods against each other, an absolute standard of comparison, namely, the exact solution of the problem of two fixed centers of gravitation, is chosen. Three lunar orbits were computed with high precision.

The reasons for this choice and the various alternatives available are discussed. Also included is a discussion of the solution of the problem of two fixed centers.

The comparison of the results leads to the conclusion that the Encke method is superior in speed and accuracy for trajectories for which the two-body problem furnishes a good local approximation (with the variation of parameters comparable as regards accuracy, but inferior with regard to speed). Cowell's method seems markedly worse than Encke's method in accuracy and speed, but is much simpler to program.

The limitations of the Encke method and the method of variation of parameters are discussed, and procedures for the removal of these difficulties are recommended.

The object of the investigation described in this paper is to make a critical comparison of the three numerical methods for orbit

---

*This report was prepared from a study carried out at Republic Aviation Corporation for the Aeronautical Research Laboratories at Wright Field [10]. The authors wish to express their appreciation for the active interest and participation of Mr. K. E. Kissell and Dr. K. G. Guderley of the Aeronautical Research Laboratories in the formulation of the problem and throughout the course of the investigation.

computation currently in common use—Cowell's method, Encke's method, and the method of variation of parameters.

Various techniques of evaluating a given numerical procedure are available. For example, Cowell's method, using a Runge-Kutta-Gill integration scheme, has been checked against the known two-body conic section solution. Comparison with the two-body solution is, however, not available for the Encke method and the method of variation of parameters, since each of these methods uses the two-body solution as the basis of a perturbation calculation. A second method is to compare the results obtained by different methods among themselves and without recourse to known standard solutions of the problem [1]. A third procedure of evaluating a given method is to introduce a check, for instance, a known integral of the motion. Thus, the invariance of the vertical component of the angular momentum of a satellite about an oblate body has been used. Such an invariant is, however, usually stable even for erroneous orbits, and thus provides a very uncritical test.

Evidently, all of the methods outlined above suffer from the lack of a standard for comparison. The problem of two fixed centers of gravitation was selected as the standard. It is integrable in closed form. An outline of the solution is given in [2], and the details appear in this paper.

The initial conditions are chosen so that the angular momentum about the line joining the two fixed centers vanishes throughout the motion. Thus, it is possible to compare the accuracy of the orbit with the degree of conservation of angular momentum.

For the comparison, three sets of initial conditions were chosen. The lowest energy orbit was a highly eccentric orbit encircling the earth, comparable to a lunar abort mission. An orbit of intermediate energy was taken describing a figure eight about the earth and moon. Finally, an orbit of high energy encircled both bodies in a single loop.

NUMERICAL   PROCEDURES[1]

Special Perturbation Methods

The equations of motion for a vehicle moving in the field of two fixed centers of gravitation may be written

---
[1]See notation, p. 46.

$$\ddot{r}_{VE} = -\mu_E \, \frac{r_{VE}}{r_{VE}^{\,3}} - \mu_M \, \frac{r_{VM}}{r_{VM}^{\,3}} \qquad (1)$$

where the subscript V refers to the vehicle, the subscript E to the earth (with respect to which the motion of the vehicle is computed), and the subscript M to the second center, the moon.

Programs for the IBM 704 were written to integrate Eq. (1) for the three special perturbation methods. Each of these was integrated using two schemes for numerical integration, a Runge-Kutta-Gill scheme and a modification of the Adams scheme using backward differences. A brief description of the different methods and integration schemes is given below. A more thorough discussion may be found in the reference cited.

1. *Cowell's Method* [3]. The Cowell method integrates the equations of motion [Eq. (1)] as they stand.

2. *Encke's Method* [3]. The Encke method uses the solution of the classical two-body problem (for the given initial conditions) as a basic solution, and integrates the deviation from this orbit by some numerical integration scheme. Let r_0 satisfy the equation

$$\ddot{r}_0 = -\mu_E \, \frac{r_0}{r^3} \qquad (2)$$

Combining Eqs. (1) and (2) the Encke equation describing the deviation of the vehicle from the two-body orbit is given by

$$\rho = -\mu_E \left[\frac{r_{VE}}{r_{VE}^{\,3}} - \frac{r_0}{r_0^{\,3}} \right] - \mu_M \, \frac{r_{VM}}{r_{VM}^{\,3}} \qquad (3)$$

3. *The Method of Variation of Parameters* [3]. This method uses as dependent variables the instantaneous values of a set of elements for a two-body orbit. The equations of motion [Eq. (1)] are transformed to give differential equations for these elements, and these are the equations that are to be integrated numerically. A wide variety of sets of parameters describing the two-body problem is available. The parameters used in this investigation are those suggested by Herrick.

The differential equations used to integrate Eq. (1) are listed without derivation.

$$\dot{A} = \frac{1}{\sqrt{\mu_E}} (r_{VE}\dot{D}' - \dot{r}_{VE}D' - FD)$$

$$\dot{B} = \frac{1}{\sqrt{\mu_E}} (r_{VE}\dot{C}' - \dot{r}_{VE}C' - FC)$$

$$\dot{a} = \frac{a^2\dot{D}'}{\sqrt{\mu_E}}$$

$$\dot{M} = n + M'$$

(4)

Integration Schemes

The two numerical integration schemes used in this investigation are a Runge-Kutta-Gill scheme of the fourth order and a scheme of the Adams type, using sixth-order backward difference. The derivation of both integration formulas are given in [4].

1. *The Runge-Kutta-Gill Scheme.* The Runge-Kutta-Gill scheme is a well-known standard integration method designed to solve a set of n first-order differential equations in n dependent variables $y_1, y_2, \ldots, y_n$, of the form

$$\dot{y}_j = f_j (t, y_k) \quad (j = 1, 2, \ldots n)$$

(5)

The differential equations of the special perturbation methods may be easily reduced to this form. The Runge-Kutta-Gill scheme approximates $f(t, y_k)$, with a third degree polynomial in time and the dependent variables in the interval t_i to t_{i+1}, and computes the integral of the polynomial at time t_{i+1}.

2. *The Adams Method.* The method of Adams is well suited to obtain the solution of second-order differential equations. The method fits a sixth-order polynomial through seven equally spaced known values of the second derivative, and integrates the extrapolated value of this derivative over the interval t_i to t_{i+1}. Let the differential equations be given by

$$\ddot{y}_j = f_j (\{y_k\}, \{\dot{y}_k\}, t)$$

(6)

The displacement and velocity are given by

$$\ddot{y}_{j,\ i+1} = \dot{y}_{j,\ i} + h \sum_{k=0}^{6} a_k \nabla^{(k)} f_{j,\ i}$$

$$y_{j,\ i+1} = y_{j,\ i} + h\dot{y}_{j,\ i} + h^2 \sum_{k=0}^{6} \beta_k \nabla^{(k)} f_{j,\ i} \tag{6a}$$

The symbol h is the uniform time interval used in the integration. The constants a_k and β_k are given in Table 1.

TABLE 1. COEFFICIENTS FOR THE ADAMS METHOD

$a_0 = 1$	$\beta_0 = 1/2$
$a_1 = 1/2$	$\beta_1 = 1/6$
$a_2 = 5/12$	$\beta_2 = 1/8$
$a_3 = 3/8$	$\beta_3 = 19/180$
$a_4 = 251/720$	$\beta_4 = 3/32$
$a_5 = 95/288$	$\beta_5 = 863/10,080$
$a_6 = 19,087/60,480$	$\beta_6 = 1925/24,192$

3. *Single and Double Precision.* In addition to testing the different integration schemes, the advantages of double-precision accumulation of the integrals versus single precision were compared. The Runge-Kutta-Gill scheme accumulated both the position and velocity in single-precision floating-point arithmetic. The Adams backward difference integration accumulated the position and velocity in double-precision floating-point arithmetic. Both schemes computed the individual increments to the solution of the equations in single-precision floating-point arithmetic.

THE PROBLEM OF TWO FIXED CENTERS

Elliptic Coordinates

In this problem one considers the motion of a third body in the gravitational field of two fixed centers of attraction. The equations of motion for the third body do not separate in rectangular coordinates. They do, however, separate in elliptic coordinates, and the orbit of the body is given in terms of elliptic functions. The solution is reduced to the evaluation of elliptic integrals in [2], for the case of planar motion, which can be guaranteed by a suitable choice of initial conditions. The initial conditions chosen for this work are shown in Figure 1. The two fixed centers μ_E and μ_M lie on the x-axis at $(\pm c, 0)$. The initial position lies on the line of centers to the right of μ_E, and the initial velocity is perpendicular to the line of centers. The symbols μ_E and μ_M refer to the gravitational constant times the masses of the two fixed centers, which represent the earth and moon, respectively.

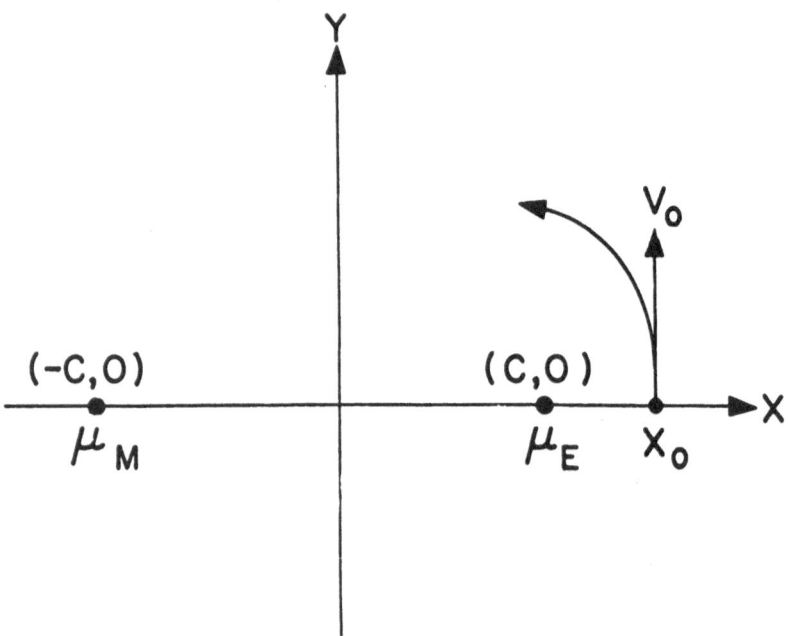

Figure 1. Coordinate system, two fixed centers.

The mathematical formulation of the problem, as well as the numerical calculation of the orbits, requires the use of elliptic functions. The three elliptic functions are sn u, cn u and dn u, and they depend not only on the variable u, but also on a parameter k, which lies between zero and one. These functions are doubly periodic with one real and one imaginary period, both depending on the parameter k [6]. Curves for one period of each are shown in Figure 2, and the following relations among the elliptic functions hold:

$$sn^2 u + cn^2 u = 1 \tag{7}$$

$$dn^2 u + k^2 sn^2 u = 1$$

with

$$sn\,(0) = 0, \qquad cn\,(0) = dn\,(0) = 1 \tag{8}$$

The relations between elliptic and Cartesian coordinates are given by

$$x = c\,\cosh\,\xi\,\cos\,\eta\,; \quad y = c\,\sinh\,\xi\,\sin\,\eta \tag{9}$$

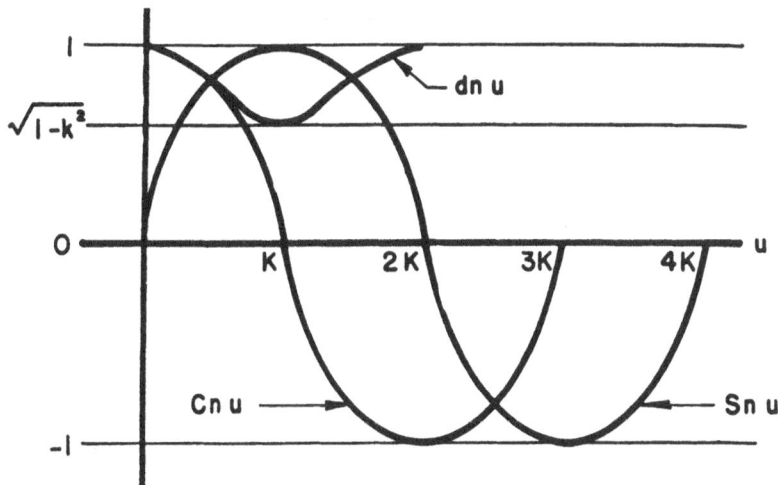

Figure 2. Elliptic functions.

In this system of coordinates, the curves ξ = const are ellipses confocal about μ_E and μ_M, and the curves η = const are confocal hyperbolas about μ_E and μ_M, as shown in Figure 3. The following special values of ξ and η may be noted:

$\xi = 0$ Segment of x-axis between μ_E and μ_M

$\eta = 0$ Portion of the x-axis to the right of μ_E

$\eta = \pi/2$ Positive y-axis   (10)

$\eta = \pi$ Portion of the x-axis to the left of μ_M

$\eta = 3\pi/2$ Negative y-axis

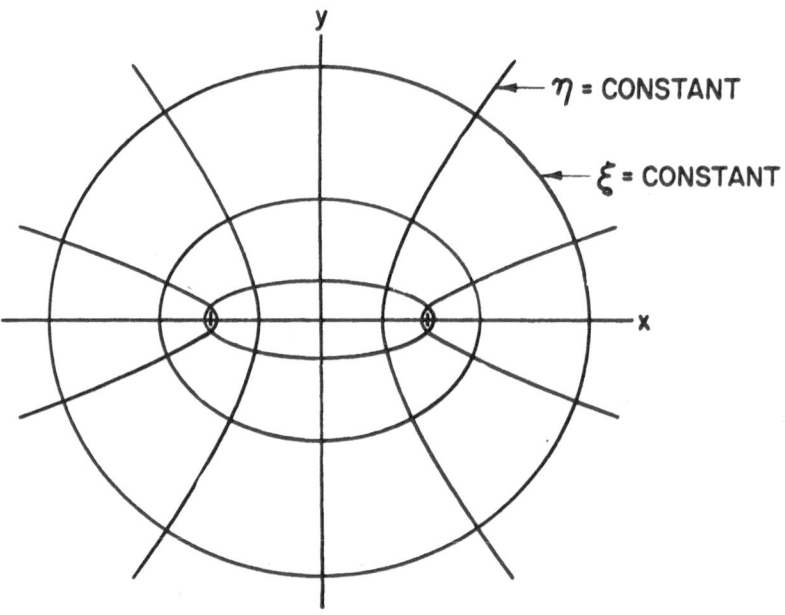

Figure 3. Elliptic coordinates.

Characterization of the Orbits

In terms of these coordinates the Lagrange equations of motion are separable and a first integral can be obtained. At this stage a parameter u is introduced, and the following differential equations remain to be integrated for ξ, η and the time t

$$du = \frac{\pm\, d\xi}{\left[h\cosh^2\xi + \dfrac{\mu_E + \mu_M}{c}\cosh\xi - p\right]^{1/2}}$$

$$= \frac{\pm\, d\eta}{\left[-h\cos^2\eta - \dfrac{\mu_M - \mu_E}{c}\cos\eta + p\right]^{1/2}} \qquad (11)$$

$$= \frac{\sqrt{2}}{c}\,\frac{dt}{\cosh^2\xi - \cos^2\eta}$$

in which h is the energy and p a constant of integration.

These equations enable one to classify the orbits into the four general types shown in Figure 4.

1. One-body orbits which encircle the mass μ_E.

2. Figure-eight orbits which enclose both bodies and cross the line of centers between them.

3. Two-body orbits which encircle both bodies, but never cross the line of centers between them.

4. Escape orbits.

These orbits correspond, for a given x_0, to increasing values of V_0. A more detailed classification for all planar orbits will be found in [5].

The equations of the orbits may be obtained from the solution of Eq. (11) for ξ and η, and the results are shown in Table 2. It will be noted that there are three possible forms for $\tanh(\xi/2)$ and four others for $\tan(\eta/2)$. The particular forms, the combination of $\tanh(\xi/2)$ and $\tan(\eta/2)$ and the values of the a's, α's, b's, and β's all depend on the values of h and p, which in turn depend on the initial conditions.

To see how the particular combinations of elliptic functions define the character of the orbit, suppose for example that $\tanh(\xi/2)$ is given by $cn\,(a_1\,u)$; then $\tanh(\xi/2)$, and hence ξ, will oscillate about zero. For $a_1\,u = K$, the quarter period of the cn function, $\xi = 0$, and the orbit crosses the x-axis between the masses. If in addition $\tan(\eta/2)$ is given by $sn\,(\beta_1\,u)$, then $\tan(\eta/2)$ will also oscillate about zero. Thus $\tan(\eta/2)$ can never become infinite; hence, η can never reach π, and hence the orbit cannot cross the

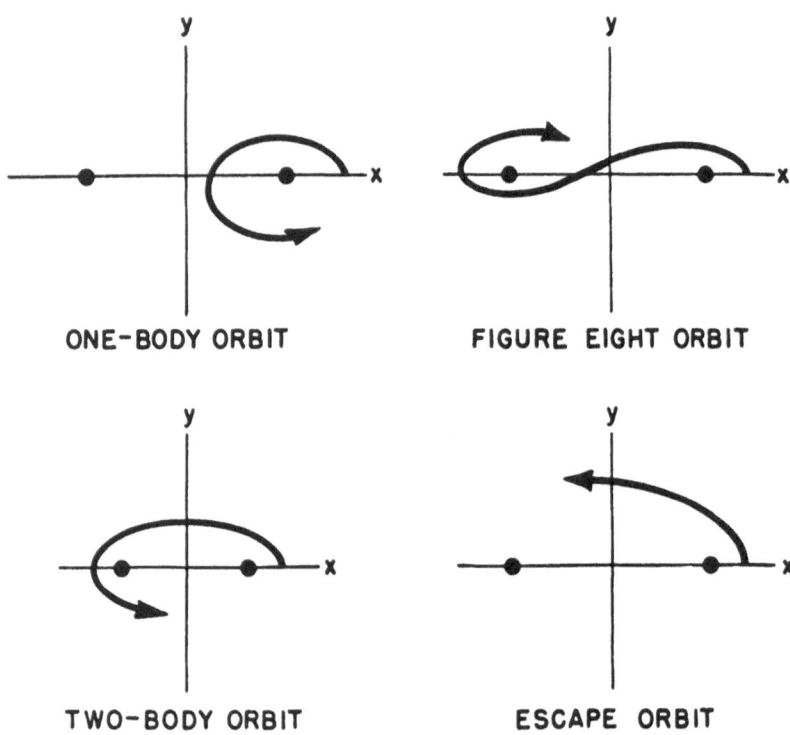

ONE-BODY ORBIT   FIGURE EIGHT ORBIT

TWO-BODY ORBIT   ESCAPE ORBIT

Figure 4. General classification of orbits for two fixed centers.

x-axis beyond the far mass. Thus, this combination leads to a one-body orbit. To consider another case, if $\tanh(\xi/2)$ is given by $1/\mathrm{dn}\,(a_3\,u)$, then the variable ξ is restricted to positive values, and hence the orbit cannot cross between the masses. Since the corresponding $\tan(\eta/2)$ is given by $\mathrm{sc}\,(\beta_3\,u)$, which can become infinite, the possibility of crossing beyond the far mass is present. Whether the orbit is a two-body orbit or an escape orbit is determined by the magnitude of the coefficient a_3. If $a_3/\mathrm{dn}\,(a_3\,u) = 1$ for any value of u, then for this value of u, ξ is infinite, and the orbit is an escape orbit. Otherwise, the orbit will be a two-body orbit.

The elliptic functions used in the solution are all periodic, but those associated with ξ will, in general, have different periods from those associated with η, and hence the orbits will be periodic only if the ξ and η periods are commensurable.

TABLE 2. ANALYTIC CHARACTERIZATION OF THE ORBITS FOR
TWO FIXED CENTERS

	$\tanh(\xi/2)$	$\tan(\eta/2)$
One-body orbits	$a_1 \, cn(a_1 u)$	$b_1 \dfrac{sn(\beta_1 u)}{dn(\beta_1 u)}$
	$a_1 \, cn(a_1 u)$	$b_2 sn(\beta_2 u)$
Figure-8 orbits	$a_1 \, cn(a_1 u)$	$b_3 \dfrac{sn(\beta_3 u)}{cn(\beta_3 u)}$
	$a_1 \, cn(a_1 u)$	$b_4 \dfrac{sn(\beta_4 u)}{cn(\beta_4 u) - b_5 sn(\beta_4 u)}$
Two-body orbits	$a_2 \, dn(a_2 u)$	$b_3 \dfrac{sn(\beta_3 u)}{cn(\beta_3 u)}$
	$\dfrac{a_3}{dn(a_3 u)}$	$b_3 \dfrac{sn(\beta_3 u)}{cn(\beta_3 u)}$
Escape orbits	$\dfrac{a_3}{dn(a_3 u)}$	$b_3 \dfrac{sn(\beta_3 u)}{cn(\beta_3 u)}$

Once ξ and η have been determined as functions of u, so also can the Cartesian coordinates x and y, and the time appears as an elliptic integral of the third kind of μ. A precision program for the IBM 704 was written to compute the coordinates and time as functions of μ. Three orbits, one of each type, were selected to serve as standards for the comparison of the special perturbation methods for integrating the equations of motion. Plots of these orbits, together with tables giving the coordinates for the first quarter period, are given in the Appendix.

Lunar Applications

The use of the problem of two fixed centers is not restricted in its suitability as a standard for comparison. It provides a surprisingly good approximation to a real lunar shot, particularly up to the first approach to the moon. In addition, it has been found that extremely good estimates as to the initial conditions necessary to achieve a close loop about the moon and specified conditions on return to the earth can be made on this model. Further, estimates can also be made on permissible deviations, both in initial conditions and conditions at any point along the trajectory in order that the specified mission may be attained. The tolerances are tight for all orbits closely circling the moon, particularly with the initial conditions. Thus, for example, the entire range of values of velocity for a figure-eight or two-body orbit varies from within 300 ft/sec of escape velocity near the earth, to a maximum of within 3,000 ft/sec of escape velocity near the moon. For a specified magnitude of velocity, the direction of the velocity must lie within two or three degrees of an angle which depends on the orientation of position relative to the line of centers.

It has been found that, for an initial position close to the earth and with sufficient velocity to attain lunar distances, as the direction of the velocity is decreased, the orbit will change as follows:

1. First, a one-body orbit about the earth and entirely above the moon.
2. Then, a teardrop orbit circling the moon and returning to the earth.
3. Then, a two-body orbit encircling both earth and moon.
4. Then, a figure-eight orbit.
5. Finally, a one-body orbit below the moon.

This sequence is shown in Figure 5.

THE COMPARISON

General Remarks

Each of the exact orbits used in the comparison lies in the plane formed by the initial velocity and the line of centers. The initial position is chosen on the line of centers. This means that the components of the angular momentum in this plane are identically zero, and in particular, the component along the line of centers vanishes.

In order to make the problem three dimensional the initial conditions used in the comparison were rotated through an angle of 30 degrees about the line of centers. This rotation made it possible to evaluate the use of an invariant of the motion as a test for the precision of the calculation. As noted above, the component of

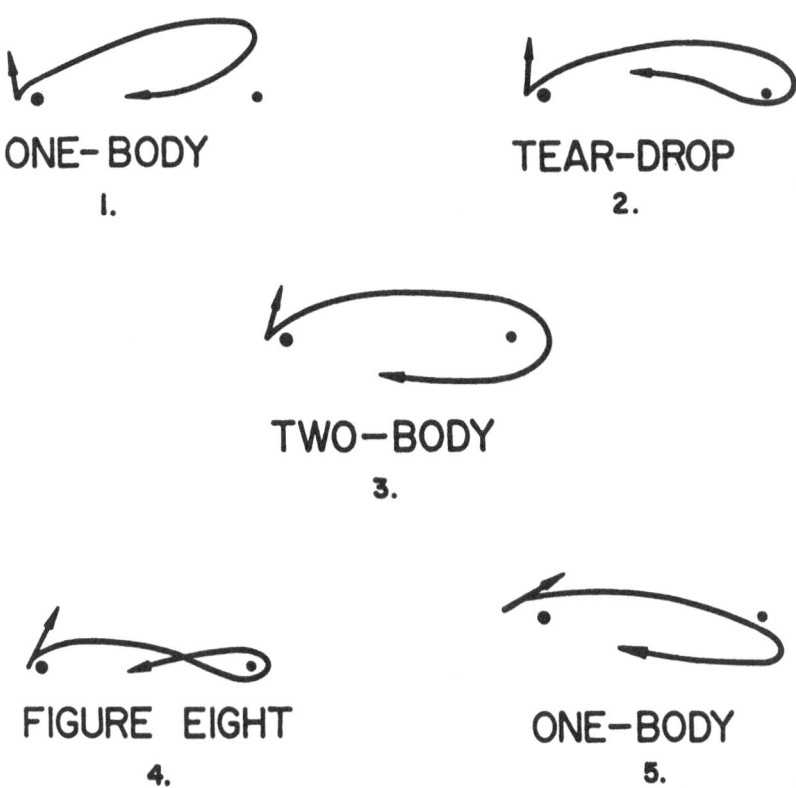

Figure 5. Variation of the orbit with direction of initial velocity.

the angular momentum about the line of centers should always vanish, so that it might be hoped that its deviations from zero would give an indication of the over-all precision of the calculation. As will be seen later, this test failed completely to give a realistic indication of the growth of the deviations from the exact orbits. The comparison of the orbits was carried out for approximately 50 hours of real time for the one-body orbit and 100 hours for the figure-eight and two-body orbits. This time proved sufficient to draw conclusions on the accumulation of errors of the various schemes.

Before proceeding, it is instructive to outline the error accumulation that might be expected, based on existing theories. A thorough investigation of the accumulation of round-off error in numerical integration schemes is contained in [7]. The basic point made in this reference is that, for the Cowell, Encke, and any of the variation of elements methods, the error accumulation in the mean longitude (or the position) builds up as the 3/2 power of the number of integration steps. On the basis of such a premise it would be expected that, for all three methods considered, the error accumulation should be proportional to the 3/2 power of the time. By plotting the absolute error on log-log paper versus the time, estimates of the exponent of the time variation of error were obtained.

Figures 6, 7, 8, and 9 contain smoothed plots of the accumulation error versus the time. Each figure contains error accumulation for the three different equations of motion, each integrated by two different integration schemes. Thus, each figure permits easy comparison of the different methods for a single orbit. The curves are labeled with two letters and a number. The first letter, C, E, or V, refers, respectively, to the Cowell, Encke or variation of parameters methods. The second letter, R or A, refers, respectively, to the Runge-Kutta-Gill or Adams integration schemes. The number gives the integration interval used. The error accumulation in y and z follows very closely that in x, so only one such curve is included.

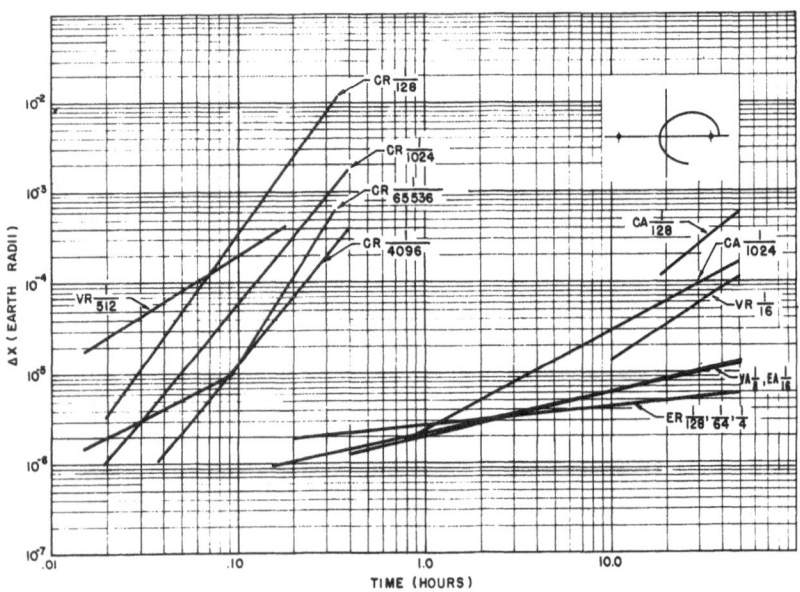

Figure 6. Absolute error accumulation in x for the one-body orbit.

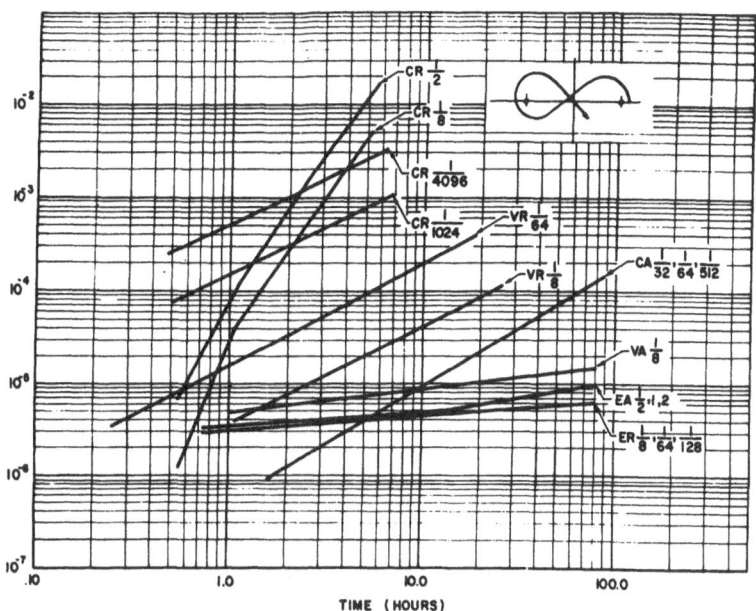

Figure 7. Absolute error accumulation in x for the figure-eight orbit.

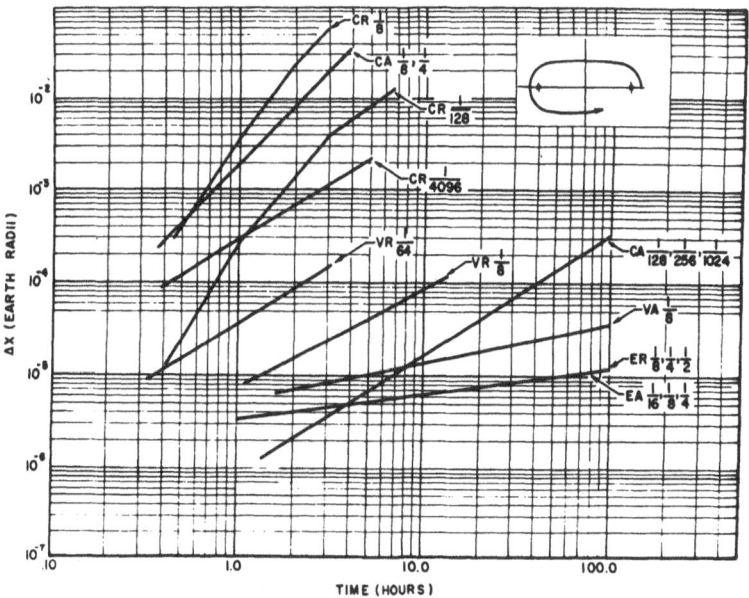

Figure 8. Absolute error accumulation in x for the two-body orbit.

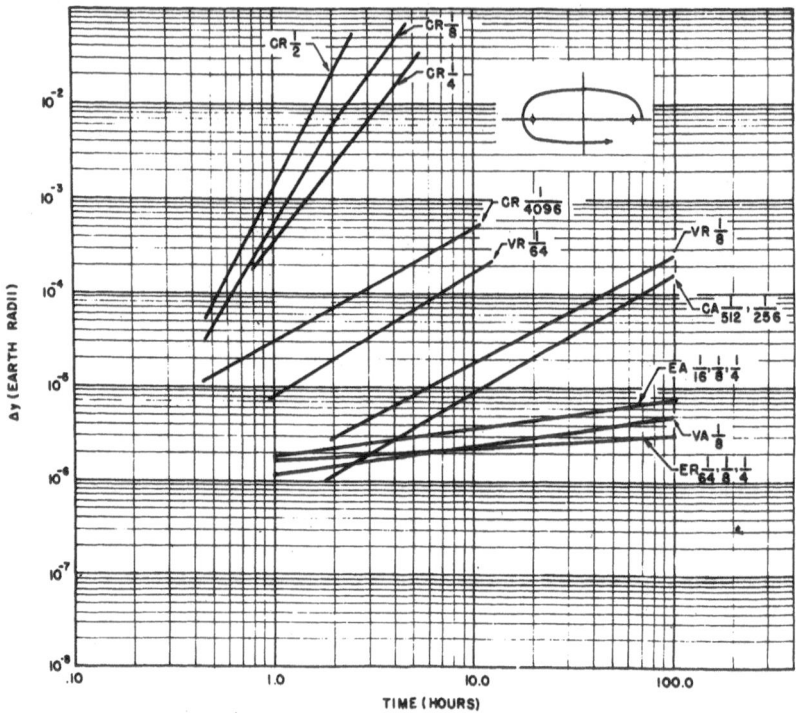

Figure 9. Absolute error accumulation in y for the two-body orbit.

Truncation Error

In carrying out the error analysis, each problem was integrated, using first a large interval size, for which truncation errors might be expected, and gradually reducing the integration interval until round-off accumulation resulted. The transition point where round-off error becomes more important than truncation is easily recognized in a log-log plot of the absolute error by the occurrence of a sudden change in error accumulation slope. The prediction of the error slope for truncation error is much more difficult to estimate than for round-off error, since it depends on the magnitude of the derivative.

The round-off error accumulation rate can be expected only if the interval size has been made small enough so that truncation errors can be neglected. An excellent discussion of the characterization of truncation errors is to be found in [4].

Figures 6, 7, 8, and 9 show the error accumulation as a function of time for the various methods with various interval sizes. It will be noted that the Cowell curves have a very steep slope for the largest intervals. In Figure 9, for example, note CR 1/2 and CR 1/8, which correspond to the Cowell method, with a Runge-Kutta-Gill integration at 1/2- and 1/8-hour intervals, respectively. As the interval size is decreased, the slope of the error curve drops to about unity. Still smaller intervals give curves with the same slope of unity, but with larger error accumulation. This is shown in Figure 9 in the curves CR 1/1024 and CR 1/4096.

Further examination of Figures 6, 7, 8, and 9 indicates that round-off accumulation occurs for the three different orbits at different interval sizes. This is to be expected, since each of the three orbits has an initial position at a different distance from the gravitational singularity of the earth. The optimum integration intervals in the round-off error region, for the three cases considered, are given for Cowell's method in Table 3.

TABLE 3. OPTIMUM INTEGRATION INTERVALS

Orbit	Initial Distance	Runge-Kutta-Gill	Adams
One-Body	1.1	2^{-16}	2^{-10}
Two-Body	5.0	2^{-14}	2^{-7}
Figure 8	12	2^{-10}	2^{-5}

Most integration schemes compute the independent variable time by accumulating the increments in t. Thus the bit representation of the time interval may become a source of error unless it can be represented as an exact binary fraction. In carrying out the integrations, serious errors in the time were noted when time intervals were used which were multiples of negative powers of ten. This error is partly eliminated in some programs by expressing the time as the product of the number of integration steps and the time interval. A superior and simpler system is to confine the integration interval size in a binary machine to negative and positive powers of two. In this way the time is computed exactly for a very large number of intervals. No difficulty was encountered in approximating a desired interval by its closest representation in the binary system.

Round-off Error

It will be noted that the curves with slope unity have been associated with round-off errors. This appears to contradict the general theory outlined in [7], from which a slope of 3/2 would be expected. This discrepancy can be explained in the following way. The machine used in the analysis was the IBM 704. This machine carries out round-off of the twenty-eighth bit by simply dropping the complete bit. Thus, a modification of the statistical theory on the accumulation of round-off errors based on zero mean of these errors is necessary.

To a first-order approximation, the increment in the velocity is given by the product of the average acceleration times the interval size.

$$\Delta \dot{x} \simeq \bar{\ddot{x}} h \tag{12}$$

The error in this value is approximately given by

$$\epsilon(\Delta \dot{x}) \simeq \bar{\ddot{x}} \, h 2^{-p-1} \tag{13}$$

where p represents the number of binary bits used in single precision computation. Since this error is monotone over a large time interval, the accumulated error is given by

$$\epsilon(\dot{x}) \cong \Sigma \, \bar{\ddot{x}} \, h 2^{-p-1} = \bar{\ddot{x}} \, 2^{-p-1} t \tag{14}$$

It is seen that the error in velocity varies linearly with time.

To the first order of approximation the increment in position is given by the product of the average velocity and the interval size.

$$\Delta x \simeq \bar{\dot{x}} h \simeq (\bar{\dot{x}}_c + \bar{\ddot{x}} \, 2^{-p-1} t) h \tag{15}$$

where the subscript c indicates the correct value of $\dot{x}$, on which the error is superimposed. The error in this increment is given by

$$\epsilon(\Delta x) \simeq (\bar{\dot{x}} + \bar{\ddot{x}} 2^{-p-1} t) \, h 2^{-p-1} \tag{16}$$

Again, since the velocity is monotone over a long period of time, the accumulated error is given by

$$\epsilon(x) \simeq \bar{\dot{x}}_c \, 2^{-p-1} t + \bar{\ddot{x}} 2^{-2p-2} \frac{t(t+1)}{2} \tag{17}$$

Examination of the accumulation error in position indicates that for an extended period, until the error proportional to the square of the time overtakes the linear error, the dominant observable error will be linear with the time. To a first order of approximation, the time period for which the first power of t will predominate over the second power of t is given by

$$T \simeq 2 \frac{\overline{x}_c}{\overline{\overline{x}}} 2^{p+1} \tag{18}$$

In floating point arithmetic in the IBM 704, $p = 27$.

In the event that a random round-off machine program is used, the same situation outlined above will prevail, except that the powers of t will be 1/2 and 3/2 instead of 1 and 2. The estimate of the time delay before the higher rate of error accumulation is noted will be the order of magnitude of the number of bits used in single-precision computation.

Conservation of Angular Momentum

The results of the tests on the deviations of the orbit from the initial plane and the lack of conservation of the x-component of angular momentum from zero are given in Tables 4 and 5 for two cases for which truncation errors build up rapidly. It will be noted that both the deviation from the plane and the x-component of angular momentum differ from zero by a quantity corresponding in magnitude to the eighth significant digit of the coordinates from which they are computed, while the errors in the coordinates accumulate to the fourth significant digit for the one-body orbit and to the second digit for the two-body orbit. It may thus be concluded that neither of these constants of the motion can provide a reliable indication of the over-all precision of the integration procedure.

Machine Time

One of the most important factors in comparing numerical integration schemes on a digital computer is the estimate of machine time necessary to compute a solution. This estimate is, of course, significant for comparison purposes, only for calculations of equal accuracy.

For the particular program used, the estimate of the machine time required to integrate over a single integration step for the various methods is given in Table 6.

TABLE 4. CORRELATION OF ERROR ACCUMULATION WITH CONSERVATION OF ANGULAR MOMENTUM FOR A ONE-BODY ORBIT

Time	Δx	Δy	Δz	$y\dot{z} - z\dot{y}$	D
0.00075096	0.0000003	0.000000001	0.00000001	0	0
0.00225289	0.0000013	0.000000021	0.00000004	0	0
0.00450601	0.0000010	−0.000000143	−0.00000024	4×10^{-9}	0
0.00600883	0.0000010	−0.000001927	−0.00000334	-2×10^{-9}	0
0.00826214	0.0000006	−0.000000834	−0.00000144	0	0
0.00976423	0	0.000001184	0.00000205	0	0
0.01201969	0.0000001	−0.000000934	−0.00000162	0	0
0.01502690	0	0.00000051	0.000000874	0	0
0.03767056	0.000018	0.00000295	0.00000510	-1×10^{-8}	0
0.06057997	−0.0000978	0.000011860	0.00002052	-2×10^{-8}	0
0.09181377	−0.0003608	0.00005284	0.00009148	0	-2×10^{-8}
0.12417942	−0.00087565	0.0001532	0.00026530	1×10^{-7}	-1×10^{-8}

TABLE 5. CORRELATION OF ERROR ACCUMULATION WITH CONSERVATION OF ANGULAR MOMENTUM FOR A TWO-BODY ORBIT

Time	Δx	Δy	Δz	$y\dot{z} - z\dot{y}$	D
0.46294974	−0.000327	−0.00003203	−0.0000555	1×10^{-7}	0
1.03658800	−0.003882	−0.0005775	−0.0010001	0	-4×10^{-8}
1.9784039	−0.021966	−0.0059234	−0.0102594	-1×10^{-7}	-9×10^{-8}
3.07543898	−0.057820	−0.0232699	−0.0403043	-3×10^{-7}	-15×10^{-8}
3.91987449	−0.089666	−0.0449717	−0.0768931	-5×10^{-7}	-9×10^{-8}

TABLE 6. MACHINE TIME PER INTEGRATION INTERVAL

	Runge-Kutta-Gill	Adams
Cowell's Method	0.074 sec	0.0575 sec
Encke's Method	0.40 sec	0.15 sec
Variation of Parameters	0.53 sec	0.18 sec

In Table 7 the interval size used and machine time required to compute the figure-eight orbit for 100 hours of real time for the three methods are given.

The time comparison for the other cases follows a similar pattern. It is apparent from Table 7 that the fastest integration scheme for a given accuracy is obtained by using the Encke-Adams method.

TABLE 7. TYPICAL INTERVAL SIZES AND MACHINE TIMES

Method	Interval Size	Machine time
Cowell-Adams	1/32 hr	3.5 min
Variation of Parameters-Adams	1/8 hr	2.5 min
Encke-Runge-Kutta	1/2 hr	1.5 min
Encke-Adams	1/2 hr	0.5 min

Double Precision

The striking discrepancies between Runge-Kutta-Gill and Adams should not be interpreted as showing the advantages of the backward-difference scheme over the Runge-Kutta-Gill scheme. The major advantage is a direct result of the accumulation in double precision used in the Adams procedure, as contrasted with the accumulation in single precision in the Runge-Kutta-Gill scheme. The addition of double precision to the accumulation of the coordinates in a Runge-Kutta scheme will greatly improve that method, so as to be comparable with the Adams backward difference. The main advantage of double-precision accumulation lies in the delay of the onset of round-off error accumulation rate proportional to t^2 as given by Eq. (18).

CONCLUSIONS AND RECOMMENDATIONS

The results of the comparison described in the preceding paragraphs show conclusively that the Encke method is superior both for precision and economy in machine time. The variation of parameters method yields comparable accuracy, but requires more machine time. Considerable refinement is possible for both of these methods, particularly for those portions of the orbits that are near-circular or near-parabolic [8, 9].

It has been pointed out that the Encke method does not show to such good advantage in problems for which a two-body solution is not a good approximation, for example, for a high thrust [1].

One might think that the Cowell method would prove superior. However, the results obtained in this investigation show that this method holds its accuracy for relatively limited periods of time even when the perturbations are small. The real answer to a precision integration of any trajectory for which Encke's method is unsuitable would be to devise a generalized Encke scheme using as a basis the closed-form solution of a related problem so selected that the deviation from the closed-form solution were small.

The principal advantage of the Cowell method is that it is simple to program. If high precision is not a requirement, this method gives reasonable accuracy for a limited time. The accuracy depends almost entirely on the integration scheme used, so that care should be exercised in its selection. It should, for example, appear mandatory that double precision accumulation be used.

The decisive advantage of the Encke method is that forces small compared to the attraction of the reference body may be accurately included in the calculation. This is, for example, not the case in the Cowell method; here all the accelerations are added, and any acceleration less than 10^{-8} times the largest acceleration will not appear in the sum. This advantage disappears, of course, if the two-body reference orbit is not locally a good approximation to the motion—that is, if large nongravitational forces such as thrust or drag are present. In such cases, the Encke method should be modified so that the reference orbit is a closed-form solution of a related problem which includes the major portion of the accelerations, and so that only small perturbations remain to be integrated numerically. The results of this investigation clearly indicate that such modifications of Encke's method are most likely to lead to high-precision techniques for trajectory analysis.

NOTATION

t	Time
r	Position vector
ρ	Perturbation of the position vector from the reference orbit in the Encke method
μ	Product of gravitational constant and mass
E, V, M	Subscripts referring to the earth, the vehicle, and the moon, respectively. Double subscripts specify relative position vectors, for example:

$$r_{VM} = r_V - r_M$$

$\nabla^{(K)}$ Backward difference operator of order K

x, y, z Cartesian coordinates of position

v Scalar velocity

c Distance between the two fixed centers

ξ, η Elliptic coordinates

h Total energy

u Basic parameter for the two fixed center orbits

k Modulus of the elliptic functions

K Quarter period of the elliptic functions

REFERENCES

1. R. Baker, et al., "Efficient Precision Orbit Computation Techniques," American Rocket Society, Preprint No. 869-59, presented at ARS Meeting, June 8-11, 1959.
2. E. T. Whittaker, *Analytical Dynamics*, New York, Dover Publications, 1944.
3. *Planetary Coordinates for the Years 1960-1980*, London, Her Majesty's Stationery Office, 1958.
4. L. Collatz, *Numerische Behandlung von Differentialgleichungen*, Berlin, Göttingen, Heidelberg, Springer-Verlag, 1955.
5. C. V. L. Charlier, *Mechanik des Himmels*, Leipzig, Veit and and Company, 1902.
6. E. T. Whittaker and G. N. Watson, *Modern Analysis*, Cambridge, Cambridge University Press, 1958.
7. D. Brouwer, "On the Accumulation of Errors in Numerical Integration," *The Astronomical Journal*, **46**:16 (1937), p. 149.
8. R. M. Leger and C. E. Herrick, "Trajectory Computation in Systems Design," Convair Astronautics Report No. AG-646 (January 1960).
9. S. Pines, "Variation of Parameters for Elliptic and Near Circular Orbits," Republic Aviation Corporation Report No. RAC-644-453 (November 1959).
10. S. Pines, M. Payne, and H. Wolf, "Comparison of Special Perturbation Methods in Celestial Mechanics," ARL Technical Report 60-281 (March 1960), Wright Field.

APPENDIX

Particular Orbits Used for the Comparison

Orbits were computed for values of μ_E and μ_M corresponding to the earth and moon, respectively. The units were earth radii, hours, and earth masses.

$$\mu_E = 19.908949$$

$$\mu_M = 0.24465681$$

$$c = 30 \text{ earth radii}$$

The initial conditions and parameters used for these three cases are given in Table 8. Plots in the x, y-plane of these orbits are shown in Figure 10 for times of approximately 160, 340, and 460 hours for the one-body, the figure-eight, and the two-body orbits, respectively. Tables 9, 10, 11 show the coordinates as functions of time for 50, 100, and 100 hours for the respective orbits.

TABLE 8. INITIAL CONDITIONS

	One-Body	Figure Eight	Two-Body
x_0	31.1	42.000005	35.000005
y_0	0	0	0
$\dot{x}_0$	0	0	0
$\dot{y}_0$	5.942	1.654	2.748

TABLE 9. COORDINATES FOR THE ONE-BODY ORBIT

Time	x	y
0.	31.100000	0.
0.15804494	30.915038	0.88910893
0.36250998	30.367466	1.7584926
0.65638871	29.478513	2.5896895
1.0763136	28.281345	3.3665991
1.6504496	26.818130	4.0762834
2.3974160	25.136621	4.7094037
3.3262475	23.286732	5.2602908
4.4371963	21.317489	5.7267047
5.7231393	19.274627	6.1093699
7.1712861	17.198926	6.4113899
8.7649429	15.125285	6.6376215
10.485140	13.082395	6.7940873
12.312019	11.092914	6.8874578
14.225907	9.1739321	6.9246367
16.208106	7.3376466	6.9124420
18.241415	5.5921000	6.8573822
20.310409	3.9419260	6.7655113
22.401581	2.3890398	6.6423463
24.503331	0.93325278	6.4928337
26.605880	−0.42719572	6.3213494
28.701143	−1.6952053	6.1317215
30.782554	−2.8744552	5.9272666
32.844898	−3.9691244	5.7108358
34.884143	−4.9836750	5.4848611
36.897266	−5.9226827	5.2514040
38.882108	−6.7907162	5.0122008
40.837240	−7.5922456	4.7687037
42.761856	−8.3315831	4.5221205
44.655650	−9.0128357	4.2734481
46.518735	−9.6398836	4.0235026
48.351578	−10.216364	3.7729464
50.154921	−10.745661	3.5223108

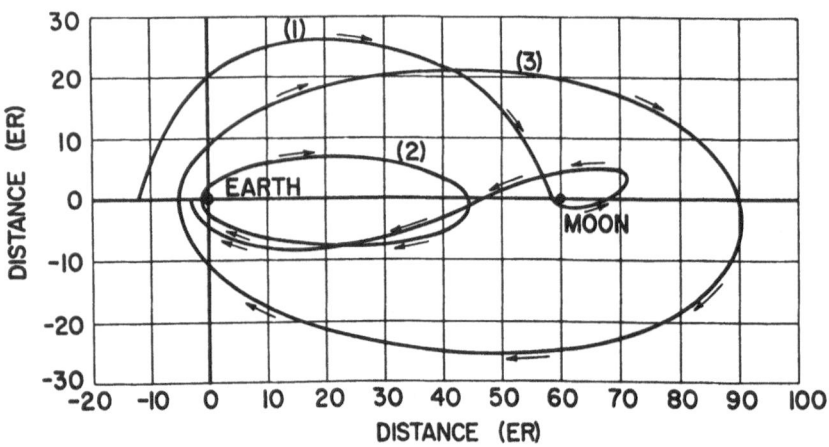

Figure 10. Exact orbits used for the comparison.

TABLE 10. COORDINATES FOR THE FIGURE-EIGHT ORBIT

Time	x	y
0.	42.000005	0.
5.4982552	40.066772	8.6205821
11.804967	34.755893	16.005343
19.337282	27.263703	21.380740
28.018762	18.929004	24.582165
37.449497	10.796828	25.890744
47.141109	3.4810826	25.776241
56.671913	− 2.7665923	24.708626
65.743332	− 7.9257366	23.071220
74.175903	−12.094654	21.142901
81.882483	−15.417312	19.111689
88.839290	−18.043295	17.096806
95.062437	−20.109110	15.169165
100.59115	−21.731400	13.367325

TABLE 11. COORDINATES FOR THE TWO-BODY ORBIT

Time	x	y
0.	35.000005	0.
0.93876813	34.663176	2.5229426
1.9784039	33.669096	4.9890792
3.2093604	32.064983	7.3461641
4.7032657	29.923290	9.5503125
6.5076487	27.333766	11.568502
8.6444714	24.394625	13.379545
11.111971	21.204322	14.973657
13.888956	17.854924	16.350978
16.940247	14.427578	17.519484
20.222296	10.990062	18.492723
23.688244	7.5961170	19.287681
27.291856	4.2860590	19.922986
30.990402	1.0881969	20.417508
34.746435	− 1.9793545	20.789373
38.528717	− 4.9067682	21.055316
42.312451	− 7.6907000	21.230328
46.079104	−10.332689	21.327501
49.815959	−12.837816	21.358024
53.515475	−15.213607	21.331263
57.174660	−17.469168	21.254899
60.794424	−19.614533	21.135087
64.379016	−21.660147	20.976622
67.935382	−23.616514	20.783101
71.472891	−25.493915	20.557062
75.002781	−27.302217	20.300125
78.537918	−29.050724	20.013107
82.092535	−30.748087	19.696119
85.681941	−32.402199	19.348656
89.322370	−34.020145	18.969675
93.030822	−35.608137	18.557662
96.824855	−37.171445	18.110694
100.72245	−38.714341	17.626502

RADIATION SHIELDING OF LUNAR SPACECRAFT

T.G. BARNES, E.M. FINKELMAN, AND A.L. BARAZOTTI

Space Sciences Group, Grumman Aircraft Engineering Corporation, Bethpage, N.Y.

A description of the radiation environment to be encountered in space is presented. Based on available experimental data, a model of the spatial, spectral, and temporal variation of charged particles from the Van Allen belts and cosmic radiation is established.

The attenuation of charged particles has been calculated using data on the range and relative energy loss as function of energy for protons and electrons in various materials. Calculations have been performed to determine the relative effectiveness of varying thicknesses of different shielding materials as a function of geometry of the shield. Shield weight penalties as a function of material and threshold energy for a representative hazard chamber are presented. Tissue dose rate variation with time for a typical three-dimensional lunar trajectory is presented and compared to flight radially in the plane of the magnetic equator. Total integrated biological dose as a function of carbon shield thickness was calculated for four solar-flare events, and for flight radially out through the inner belt. Considering the amounts of ablation material, structure, and equipment surrounding the crew in a reentry vehicle, it was found that the additional shielding required for traversal of the belts is small. It is shown that approximately 10 g/cm^2 of carbon shielding is sufficient to shield not only the inner-belt protons, but would also provide adequate protection against solar flares of the magnitude that occurred in March and August of 1958. However, it was also found to be impossible to shield against solar flares of the highest known energy within the reasonable weight limits of a typical lunar mission. Since adequate shielding cannot be provided against these very intense solar proton events, the probability of encountering these and and various lesser intensity flares is extremely important. Accordingly, the probability of encountering solar protons as a function of mission time is presented. These encounter probabilities are considerably reduced if flares can be predicted. Continuing effort in this area indicates that some degree of solar-flare prediction is feasible.

During the last several years, the developments in radiation measurement instrumentation, earth satellites, space probes, and balloon techniques have provided knowledge that there exists, above the earth's atmosphere, particle radiation which can be both biologically hazardous and damaging to sensitive spacecraft equipment. The intent of this paper is to review what is presently known of this environment, and to discuss its implications for unmanned and manned cislunar flight, from the standpoint of the spacecraft designer. Only galactic cosmic rays, solar protons, and both protons and electrons trapped in the earth's magnetic field are considered. Although the state of our knowledge on these radiations has been more completely summarized elsewhere [1-4], they are briefly repeated below for the sake of completeness.

Galactic Cosmic Rays

The galactic background of cosmic rays has been studied for many years at the top of the earth's atmosphere with high-altitude balloons and rockets. From these measurements, and using the earth's magnetic field as a momentum selector, the energy spectrum and the particle composition of the primary radiation is now quite well known [5]. These particles consist of about 85 per cent protons, 13 per cent alpha particles, and less than 2 per cent higher atomic number particles, all stripped of their electrons and moving isotropically at very high velocities. The remarkable thing about galactic cosmic rays is that the energy spectrum extends up to at least 10^{12} mev. These great energies are thought to arise from a slow acceleration in the galaxy, perhaps by a mechanism involving the collision of charged particles with moving magnetic clouds.

Therefore, these particles are not readily attenuated by matter, and small amounts of shielding may even increase the dose to space crewmen. The energy degrades chiefly by complicated nucleonic processes involving the production of mesons, nuclear evaporations, and electronic and photonic cascades. That these primary particles do not reach the earth is due to atmospheric attenuation, which is roughly equivalent to 10 m of water. Only a small fraction of the secondary particles produced are detected at the earth's surface. Recent measurements by the Pioneer V space probe [5] place the free-space of flux of particles at $2.5/cm^2$-sec and an ionization rate inside at approximately 1 g/cm^2 of low atomic number shielding material of 0.6 mr/hour. The total integrated flux and dose for a year may then be less than 10^8 protons/cm^2 and 10 r. Thus, galactic cosmic ray particles alone will only present a hazard to manned orbiting space stations above the protection of the earth's magnetic field, or to manned lunar bases, where durations greater than a year are anticipated.

Radiation Belts

The model of the radiation trapped by the earth's magnetic field that is in general use was proposed by Van Allen [6], and is illustrated in Figure 1. Shown are the contours of true counting rates that were constructed from instrumentation flown in Explorer IV, Pioneer III, and Pioneer IV. Van Allen has suggested that the heart of the inner and outer zones consists of the following:

Inner Belt:

protons, $E > 40$ mev	$\sim 2 \times 10^4/\text{cm}^2\text{-sec}$
electrons, $E > 20$ kev	$\sim 2 \times 10^9/\text{cm}^2\text{-sec-ster}$
$E > 600$ kev	$\sim 10^7/\text{cm}^2\text{-sec-ster}$

Outer Belt:

protons, $E > 60$ mev	$\lesssim 10^2/\text{cm}^2\text{-sec}$
electrons, $E > 20$ kev	$\sim 10^{11}/\text{cm}^2\text{-sec}$
$E > 200$ kev	$\lesssim 10^8/\text{cm}^2\text{-sec}$
$E > 2.5$ mev	$\lesssim 10^6/\text{cm}^2\text{-sec}$

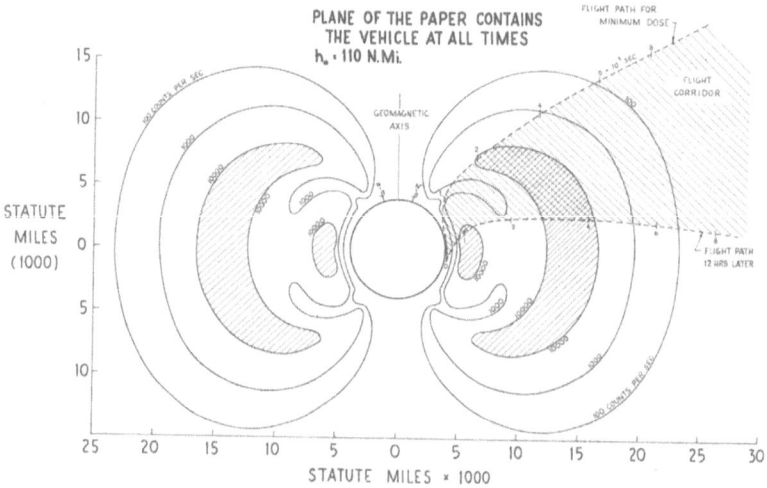

Figure 1. Three-dimensional trajectory through the Van Allen belt.

Complete data on energy spectra of particles as a function of position and time are yet to be measured. It is believed that the proton content of the inner zone is relatively stable in location and intensity, and may not vary greatly with time. It is now established that these protons are the result of neutron albedo from the collision of cosmic rays with the earth's atmosphere. However, the outer belt has exhibited fluctuations in intensity and position with time, evidently being affected by magnetic storms accompanying solar-flare events [7,8]. The origin of the electrons in the outer belt is still not established.

The lower regions of the inner zone have been probed with a nuclear emulsion package boosted by a Thor-Able vehicle [9]. The proton spectrum that was measured is shown in Figure 2. The attenuated curve for 0.1 in. aluminum will be discussed later. Although there is no reason to believe that this primary spectrum holds throughout the inner region, it has been assumed to apply for lack of other data. Fortunately, the NASA-sponsored NERV program of probing this region with additional emulsion packages should soon make additional data available.

For the purposes of calculation, an analytical fit was made to the integral electron fluxes at the heart of the outer belt that were suggested by Van Allen, to arrive at a differential energy spectrum of the form,

$$N(E)dE = (A_1 E^{-n_1} + A_2 E^{-n_2})\ dE$$

The resulting spectrum is shown in Figure 3. In view of the small amount of experimental data available, the above form of the outer-belt electron spectrum should be considered only as speculative. However, it is interesting that the slope of this spectrum approximates that of a spectrum that was measured in the edge of the outer belt at high latitudes from 600 to 1,000 km altitude [4], also shown in Figure 3.

Solar Flares

The greatest potential radiation hazard for cislunar flight will come from solar flares, as discussed later. It is now known that the surface of the visual disc of the sun frequently erupts into flares which can result in gusts of energetic protons reaching the vicinity of the earth [2,5]. These clouds of almost pure protons,

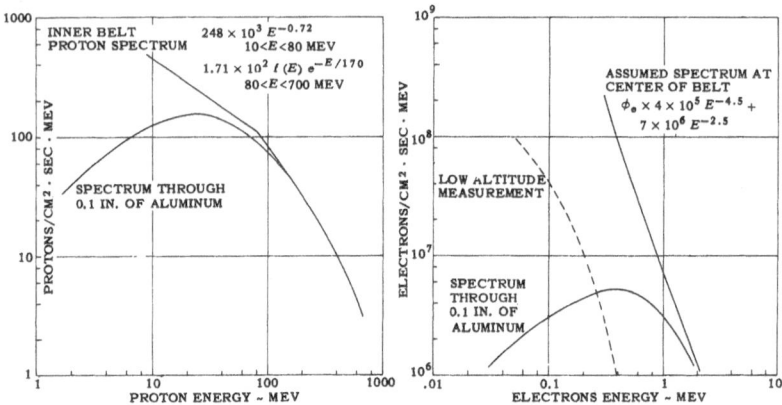

Figure 2. Inner-belt proton data.   Figure 3. Outer-belt electron data.

which have at times increased the free-space cosmic ray counting rate by a factor of 1,000, rush into the atmosphere along the lines of force of the earth's magnetic field. It is generally believed that this proton flux is omnidirectional in the vicinity of the earth, although it is hoped that it may appear more collimated in the first few hours after arrival. This is yet to be determined.

Anderson [10] has illustrated the order of events in a typical flare, reproduced here as Figure 4. The figure shows a composite picture of several actual flares, and it should be emphasized that every flare does not follow exactly this sequence. The flare light, ultraviolet, and X-radiations reach the earth with the speed of light, and may last for four hours. Concurrently, high- and low-frequency solar radio emissions may be received. Within several hours, protons begin arriving at the earth, and may continue for two days. The background ratio noise of our galaxy, as measured by rheometers near the poles, drops during the event as a result of increased ionization in the ionosphere, due to the arrival of ultraviolet, X-rays, and then protons. Late-arriving clouds of low-energy electrons and protons then disturb the earth's magnetic field, which may allow penetration of solar protons at lower, normally "forbidden" latitudes.

Solar-flare emissions of cosmic rays are usually divided into two categories; the first are very large flares emitting very high energy particles, of which there have been six in the last eighteen years; the second are those discovered during the IGY, and are characterized by higher-intensity fluxes of protons of lower, non-

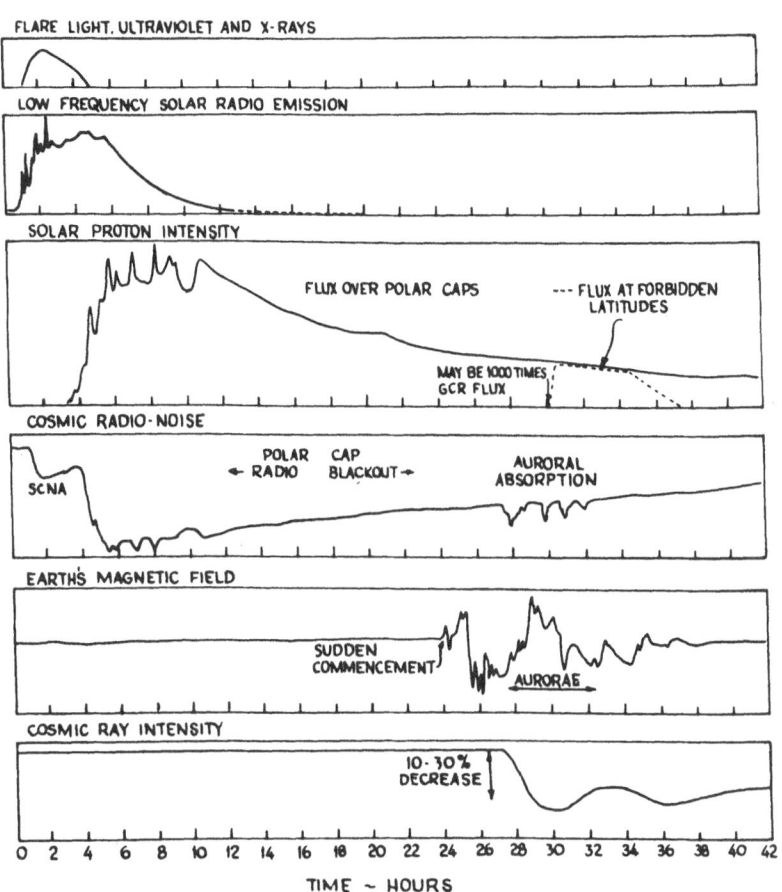

Figure 4. Solar-flare phenomena.

relativistic energies. Both result from large optical flares that emit much radio noise. The frequency of the lower-energy flares was approximately one per month during the last period of maximum activity in the solar cycle. Even within this group of flares, the variation in duration and intensity covers a wide range. During the flares of March 1958, August 1958, May 1959, and July 1959, particle measurements were recorded by balloons and satellites. These are summarized in Table 1. Shown on this table are the date and time of arrival of protons as given by the "blackouts" in polar-cap cosmic radio-noise absorption [11], and the significant data measured. Data from several events in 1960, as measured by the NASA/GSFC

TABLE 1. AVAILABLE SOLAR PROTON MEASUREMENTS

Flare	Polar Cap Absorption	Significant Data	References
3/23/58 0950 UT	3/25/58 by 2230 UT	First balloon measurement of solar protons. Flight began 1400, 3/26, lasted four (4) hours, 0.11 protons/cm^2-secster, at 1810 UT, above 120 mev.	J. R. Winckler, May 1960 *Journal of Geophysical Research*
8/22/58 1417 UT	8/22/58 by 1700 UT	Balloon aloft when protons arrived at 1530 UT, 8/22, at Churchill. $N(E) \, dE = K(t)E^{-5} dE$ from ascent data. Measured down to 100 mev. Believe can extrapolate spectrum down to 30 mev. Satellite 1958 E measured 100-570 protons/cm^2-sec at 0930 8/23, 50° N. latitude. Suggest spectrum $N(>E) = 5 \times 10^8 E^{-4}$/cm^2-sec at 0430, 8/23.	K. A. Anderson *et al.*, September 1960 *Journal of Geophysical Research*
5/10/59 2000 UT	5/11/59 0130 UT	Placed balloon aloft about 30 hours after arrival of protons, measured flux — 1,000 protons/cm^2—sec above 110 mev, $N(E) \, dE = KE^{-4.8} \, dE$, between 110-220 mev.	J. R. Winckler, *Physical Review Letters*, August 15, 1959
7/14/59 0342 UT	7/14/59 0700 UT	From measured flux, inferred free space flux of 1.76×10^4 protons/cm^2—sec 29 hours after flare, or possible 1.5×10^7 1 hour after flare, all for 40-500 mev.	J. R. Winckler Paper presented to Radiation Research Society, May 10, 1960 Also, R.R. Brown *Physical Review Letters*, October 15, 1959

sounding-rocket program at Fort Churchill, Canada, were not available at this writing. The differential-energy spectrum that has been used was measured by Ney and others [12] for the May 1959 event, and this is reproduced in Figure 5.

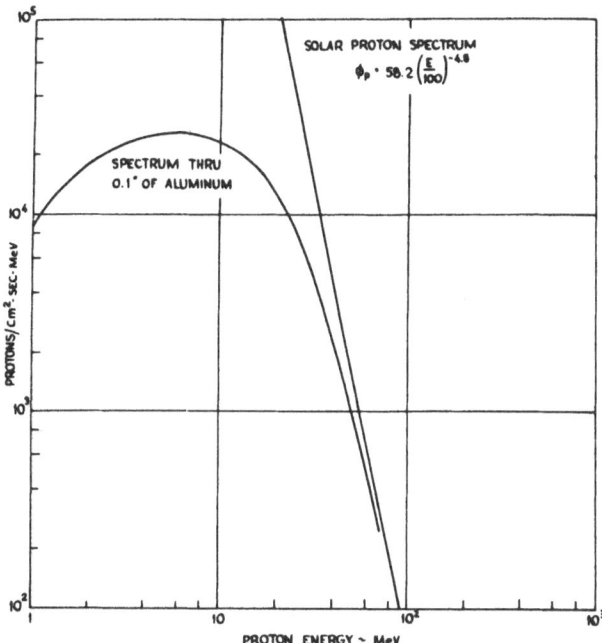

Figure 5. Solar proton data (May 1959 flare) 33 hr after observation of flare.

In order to estimate the total integrated solar proton flux that a spacecraft may experience, it is necessary to assume a model for these events. Calculations at Grumman have been based on the following time profile. The proton intensity in the vicinity of the earth grows to its maximum value for one hour following the arrival of protons, as signaled by absorption of cosmic radio noise. It remains at this maximum value for three hours and then decays as t^{-2}, as suggested by Winckler [5]. Figure 6 shows the resulting time profiles for these flares, drawn through the measured data points. It should be emphasized that the above model was assumed for the purpose of obtaining order-of-magnitude answers, and that such models will be improved as new data are available.

Shielding Calculations and Materials

The shielding analysis of charged particles can be divided into (1) attenuation of primary particles, and (2) the production and attenuation of secondary radiation, within the shield or spacecraft

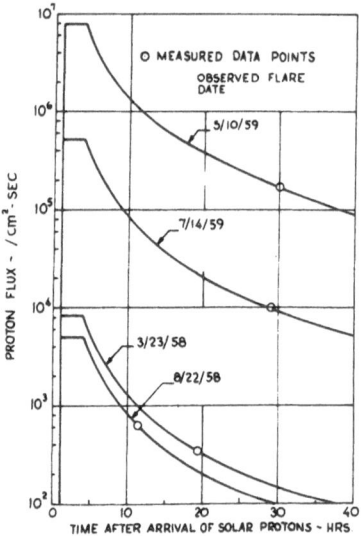

Figure 6. Solar proton events.

structure. The attenuation of primary protons and electrons in shields has been calculated, using available data on range and relative energy loss (REL) in various materials. The following assumptions were made in the calculations.

$$-\frac{1}{\rho}\frac{dE}{dX} = AE^{-N}$$

$$G(E)dE = CE^{-n}dE, \quad E_i < E < E_m$$

where

$-\frac{1}{\rho}\frac{dE}{dX}$ = REL in shield material

ρ = density of shield material

$G(E)dE$ = incident differential charged particle energy spectrum

E_m = maximum energy particle in source term

E_i = lowest energy particle in source term

The resulting differential-energy spectrum through the shield, $G(E')dE'$, is

$$G(E')dE' = C[E'^{(N+1)} + (N+1) A\rho t]^{-(n+N)/(N+1)}E'^N dE'$$

where t = shield thickness and A, C, N, and n are constants.

The estimated omnidirectional fluxes were assumed to be normally incident on a slab shield for purposes of calculating dose rates in a spherically shaped crew compartment. The equation used for biological dose rate from primary protons was

$$\text{Dose Rate} = K\int_{E_i'}^{E_m'} R(E')G(E')I(E')dE'$$

where $E_m' = [E_m^{(N+1)} - (N+1)A\rho X]^{1/(N+1)}$

$$E_i' = [E_i^{(N+1)} - (N+1)A\rho X]^{1/(N+1)},$$

$$\text{for } E_i^{(N+1)} > (N+1)A\rho X$$

$$= 0, \text{ for } E_i^{(N+1)} \leq (N+1)A\rho X$$

K, $R(E')$, $I(E')$ = a constant including normalization and conversion factors; the RBE for charged particles at energy E'; and the REL in tissue for energy E'

Similarly, the differential energy spectra emerging from 0.1-in. of aluminum were determined for the trapped radiation and solar protons, as shown in Figures 2, 3, and 5. As discussed later, these calculations were made to obtain an estimate of the radiation damage possible to sensitive equipment inside spacecraft structure.

The production of secondary radiation due to proton interactions with matter represents a problem area that must be more thoroughly investigated before shields can confidently be designed. Spallation and evaporation of nuclei by high-energy protons result in (p, xn) reactions. Coulomb excitation of atoms by protons of lower energies will produce secondary photons. Unfortunately, there is a scarcity of cross-section data for these reactions at the present time. Preliminary calculations for the reaction C^{12} $(p, pn)C^{11}$ indicate that secondary neutrons can make a significant contribution to the total dose rate through a shield. The production of secondary *Bremsstrahlung* X-radiation from decelerating electrons was also calculated, using photon spectral distributions suggested in [13].

The choice of shield materials (for minimum weight) depends upon the radiation types and geometry. Considering the attenuation of primary protons above, Figure 7 illustrates that low atomic weight materials would be the lightest in slab geometry. Figure 8 indicates that the effect of density tends to favor heavier materials when spherical or cylindrical shielding is required, and as the radius of the curved surface is decreased. This latter figure shows the shield weight per unit of inner surface area for stopping protons of energies less than 240 mev as a function of inner spherical radius. Note that carbon is a promising choice over a reasonable range of radii. However, the situation is complicated by the fact that carbon has a high cross section for the production of secondary neutrons, requiring a more thorough analysis. The *Bremsstrahlung* from electrons is best attenuated by high atomic weight materials.

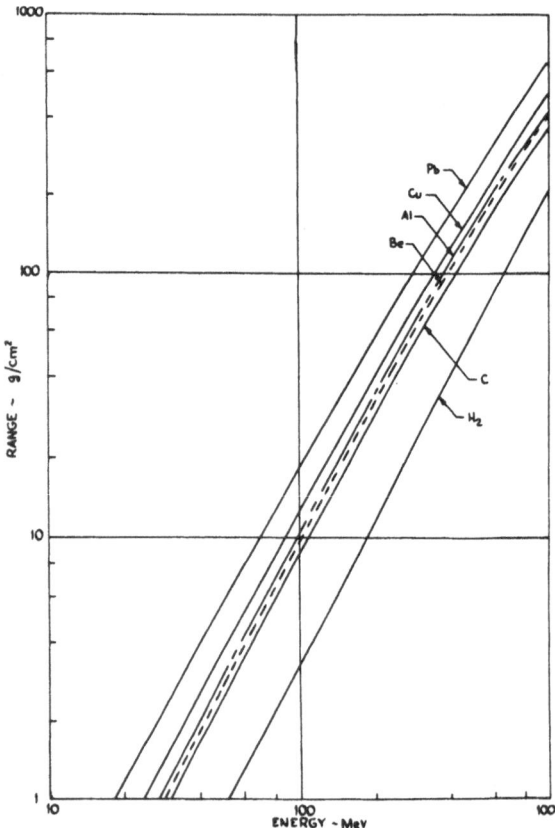

Figure 7.

Range-energy relations for protons.

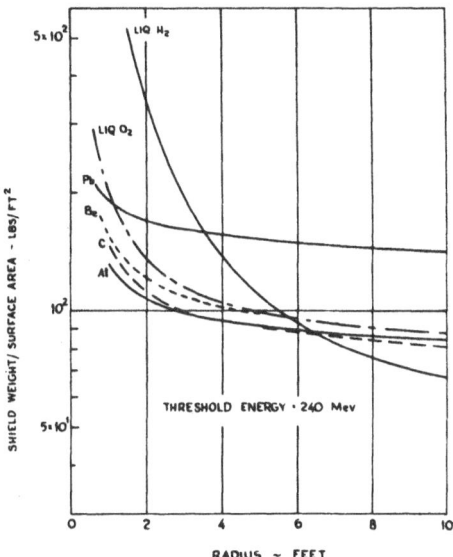

Figure 8. Proton shield material comparison.

Implications to Cislunar Spacecraft

The effects of the radiation belts were determined in the following manner. Figure 1 illustrates a possible corridor of flight paths
from near-earth orbits to the moon. This corridor was established
by a three-dimensional trajectory calculation, starting from about a
100-mile orbit, by varying the time of day of launch such that the
flight path appears to rotate ±17 degrees with respect to the belts.
For this case, it appears that it may be possible to avoid the maximum intensity region of the inner belt on a single pass, but not the
outer belt. Assuming that the flight passed directly through the
center of the belts, the vehicle would be in the heart of the inner
and outer belts for about 10 and 20 minutes, respectively. To estimate the integrated fluxes for traversal of the belts, it was assumed
that the particle flux at any position was equal to the fluxes in the
heart of the zones, as given earlier, multiplied by the ratio of the
counts per second that would exist at that position, to the maximum
counts per second in the heart of each zone, taken as 25,000. The
results obtained are believed typical for lunar flights leaving near-
earth orbits with high acceleration.

For unmanned spacecraft, it is expected that solid-state electronic components, particularly solar cells, will be among the most
radiation-sensitive items.   Experimental radiation-damage data
[14-17] for silicon solar cells are shown in Figure 9. Brown has
suggested that recent data may lower the integrated proton flux
shown in this figure by a factor of 2. Using the foregoing trajectory data, the integrated electron and proton fluxes for a single
traversal of the belts are compared to the solar-cell-damage data in
Table 2. Also included is the estimated integrated proton flux for
the solar flare of May 1959.

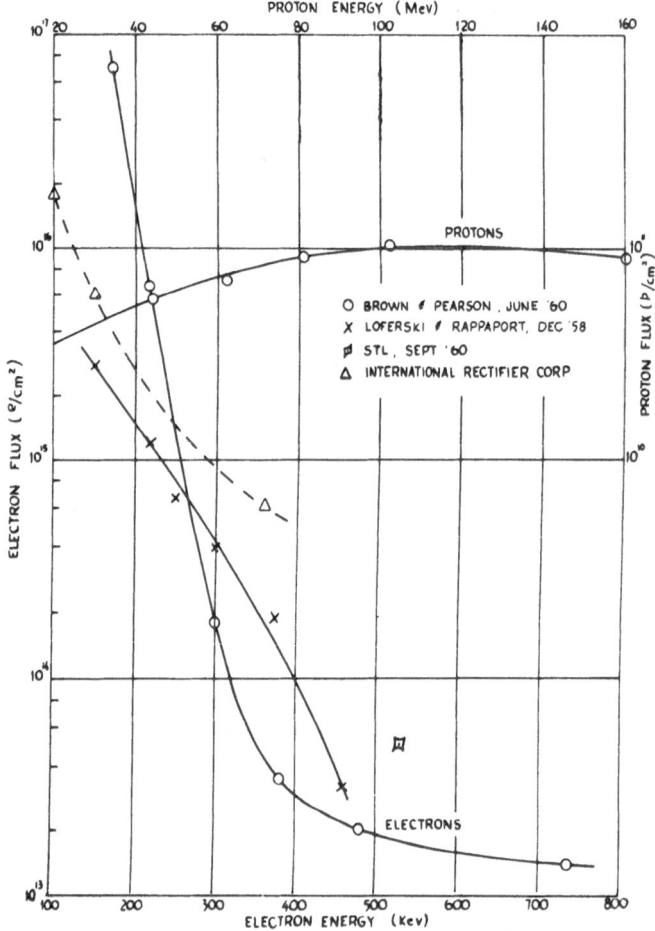

Figure 9. Solar cell damage data — Integrated fluxes for 25% reduction in
power.

TABLE 2

Source	Energy Range	Incident Flux	Flux for Damage
Inner Belt Protons	20-60 mev	$5 \times 10^6 \, p/cm^2$	$5 \times 10^{10} \, p/cm^2$
Outer Belt Electrons	0.2-0.6 mev	$1 \times 10^{11} e/cm^2$	$1 \times 10^{15} e/cm^2$
	0.6-1.0 mev	$1 \times 10^{10} e/cm^2$	$2 \times 10^{13} e/cm^2$
	1.0-1.4 mev	$2 \times 10^9 e/cm^2$	$1 \times 10^{13} e/cm^2$
Protons from Solar Flare	35-140 mev	$1 \times 10^{11} p/cm^2$	$8 \times 10^{10} p/cm^2$

The accumulated dose possible within an unmanned vehicle has also been estimated, using 0.1-in. thickness of aluminum as representative of vehicle skin and equipment casing. The dose absorbed by a silicon transistor in the vehicle was calculated, assuming no additional protection from surrounding equipment. As above, the contribution from the belts was small compared to a large solar flare. The dose absorbed during one traversal of the belts amounted to less than 10 rad for the inner belt and about 100 rad for the outer belt. The dose deposited by the flare was of the order of 10^5 rad. This assumes that the solar proton spectrum can be extended down to 20 mev. Assuming equal-energy-deposited results in equal damage for both charged and uncharged particles, this dose approaches the threshold of damage for transistors and diodes.

Therefore, the principal source of radiation damage confronting the unmanned lunar spacecraft will be solar protons. While the addition of coverglass over solar cells will appreciably reduce damage from electrons trapped in the radiation belts, it affords little protection against either inner-belt or solar protons. A more promising approach appears to be in the development of more radiation-resistant cells, such as the n-on-p type. Earth satellites with highly eccentric orbits that extend beyond the protection of the earth's magnetic field also face this problem. Damage to sensitive items within the spacecraft may be reduced by equipment layouts that provide some degree of self-shielding.

Considering now manned spacecraft, the tissue dose rates that would be experienced by a spacecraft passing through the center of the three-dimensional trajectory corridor (Figure 1) and by passing radially out along the plane of the magnetic equator are shown in Figures 10 and 11, respectively. The total integrated biological

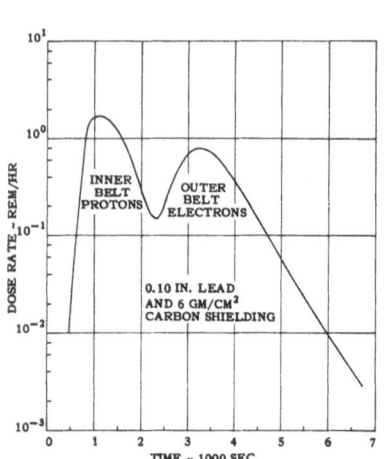

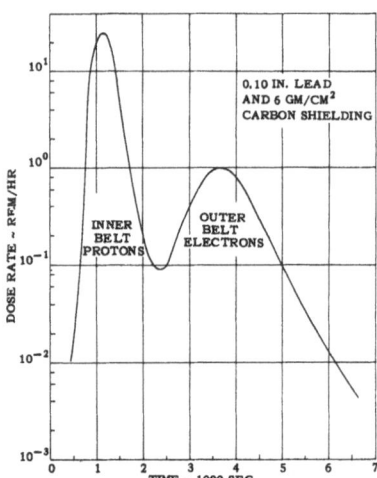

Figure 10. Tissue-dose rate for three-dimensional trajectory lunar flight.

Figure 11. Tissue-dose rate for equatorial lunar flight.

dose from primary protons has been calculated as a function of carbon-shield thickness for the four solar flares listed in Table 1, and for a flight radially out through the inner belt (Figure 12). Also shown are possible dose limits of 5 rem for the design case, and 25 rem for emergency. Approximately 10 g/cm^2 of carbon shielding would adequately shield against inner-belt protons, and also provide sufficient protection against flares of the magnitude of March and August 1958. However, the flare of May 1959 would require over 40 g/cm^2 of carbon. Tissue doses were calculated assuming RBE's of 2.0 for proton energies less than 40 mev, and 1.0 for higher energies, as suggested by Schaeffer [18, 19].

For the Van Allen belts, the shield thicknesses required to attenuate primary protons far exceed the maximum range of the more energetic electrons. However, it was found that the *Bremsstrahlung* requires that the inside of crew compartments be lined with about 0.1 in. of lead, or equivalent thicknesses of other high atomic weight material.

A rough estimate of the minimum weight of a one-man shield is shown in Figure 13, from which it is seen that the May 1959 flare would require a shield of over 5,000 lb. Considering that manned lunar spacecraft will have a crew of at least two, the significance

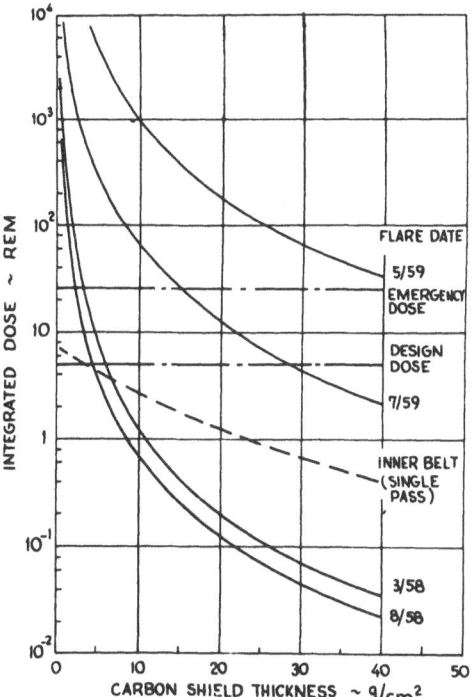

Figure 12. Proton shielding.

of solar protons becomes apparent. It is obvious that radiation protection now becomes one of the most influential factors in the selection of spacecraft configuration. Maximum use of ablative materials, propellants, equipment, and structure must be made to reduce the undesirable dead weight of shielding. Figure 14 illustrates one such configuration, wherein a hazard compartment is provided within a liquid oxygen propellant tank.

Since solar flares are potentially the greatest radiation hazard to lunar spacecraft, the next question is encounter probability, and how this probability may vary with flare intensity. The very high energy events, which emit protons with relativistic velocities, and which have occurred once every three years on the average, would be much more damaging than those considered in this study. The flares listed in Table 1 may occur about once per month during solar maximum. Anderson [20] suggests that this may reduce to less than three per year at solar minimum. Anderson further sug-

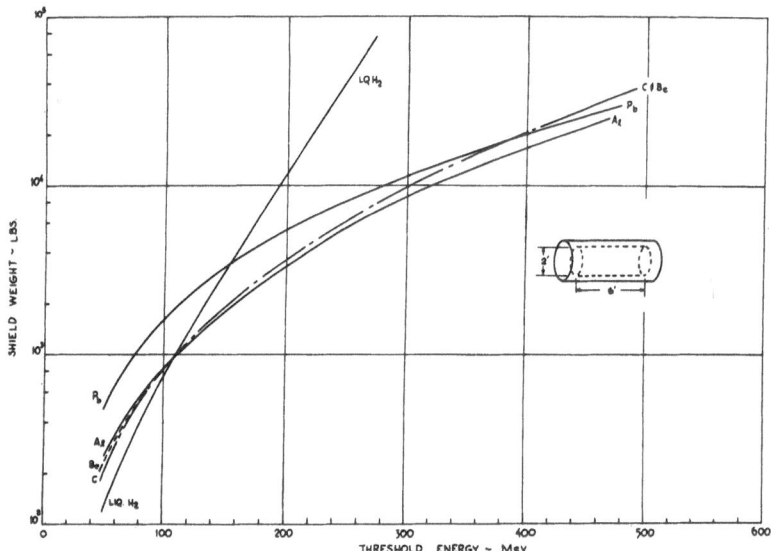

Figure 13. Estimated weight of a one-man shield.

gests that even at solar maximum the rate of occurrence of events such as May 1959 may not exceed one per year. These extremes of encounter probability are shown in Figure 15 as a function of mission time. Low-energy, "swing-around", round-trip missions to the moon require about seven days. It is doubtful that round trips can easily be reduced below five days.

These probabilities can be improved if flares can be predicted. Fortunately, recent work by Anderson [21] for the NASA has indicated that some flare prediction may be feasible. By studying over 1,200 detailed drawings of the sun's surface that had been made daily at the Athens Observatory in Greece for the last three years, Dr. Anderson was able to identify distinguishing characteristics of growing sun-spot groups that eventually emitted flares. He was able to develop a criterion which was effective in anticipating flares by at least two days in nearly all cases. The use of this criterion would have reduced the available operating time by about two-thirds during the last period of solar maximum. It is doubtful that this method can be used to extend predictability beyond four days, although efforts are continuing in this direction. If predictability cannot be extended beyond several days, manned operations

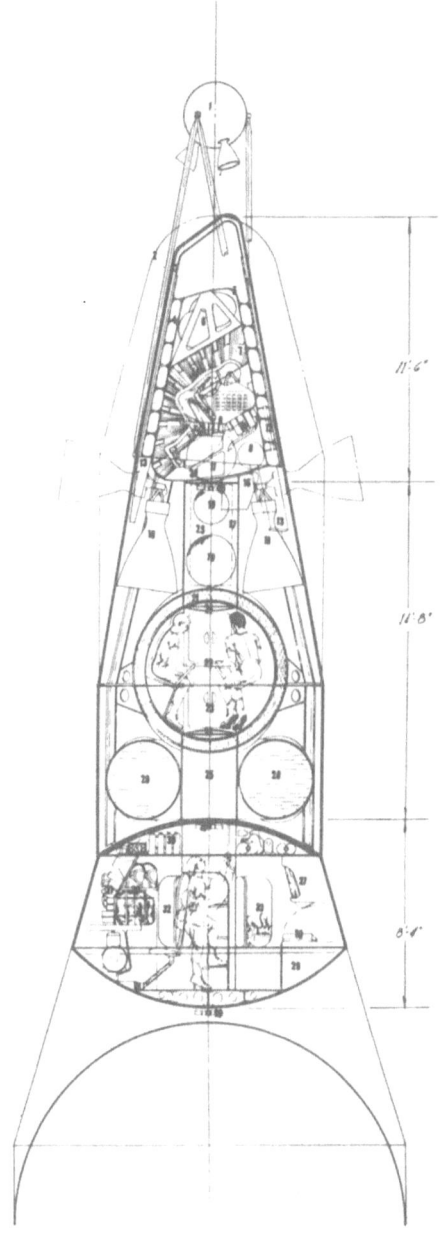

Figure 14. A manned lunar spacecraft.

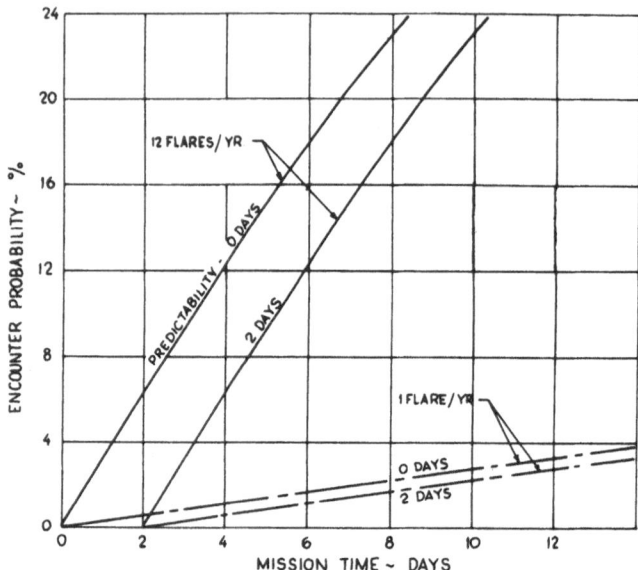

Figure 15. Solar-flare encounter probabilities.

beyond the protection of the earth's magnetic field will be a matter of gaming with solar proton encounter probabilities.

For the future, further data describing the trapped radiation and the statistics of solar proton events will be required, particularly for manned lunar flight. In this respect, the results of NASA planned experiments for measuring solar-beam particles will be significant. Better calculations on the relative contribution of secondary radiation produced in shields must also be made.

REFERENCES

1. Robert Jastrow (ed.) "Symposium on the Exploration of Space," *Journal of Geophysical Research,* **64**:11 (November 1959).
2. J. R. Winckler, "Balloon Study of High Altitude Radiations during the International Geophysical Year," *Journal of Geophysical Research,* **65**:5 (May 1960).
3. B. Peters, "Progress in Cosmic Ray Research since 1947," *Journal of Geophysical Research,* **64**:2 (February 1959).
4. Bruno Rossi, "Scientific Results of Experiments in Space," *Transactions, American Geophysical Union,* **41**:3 (September 1960).
5. J.R. Winckler, "Primary Cosmic Rays," a paper presented at the Radiation Research Society Symposium, May 10, 1960.

6. J.A. Van Allen, and L.A. Frank, "Radiation Measurements to 658, 300 km with Pioneer IV," State Univ. of Iowa Report SUI-59-18 (August 1959).

7. R.L. Arnoldy, R.A. Hoffman, and J.R. Winckler, "Observations of the Van Allen Radiation Regions during August and September 1959, Part I," *Journal of Geophysical Research,* **64**:11 (November 1959).

8. P. Rothwell and C.E. McIlwain, "Magnetic Storms and the Van Allen Radiation Belts—Observations from Explorer IV," *Journal of Geophysical Research,* **65**:3 (March 1960).

9. S.C. Freden, and R.S. White, "Particle Fluxes in the Inner Radiation Belt," *Journal of Geophysical Research,* **65**:5 (May 1960).

10. K.A. Anderson, "Solar Particles and Cosmic Rays," *Scientific American,* **202**:6 (June 1960).

11. G.C. Ried, and H. Leinbach, "Low Energy Cosmic-Ray Events Associated with Solar Flares," *Journal of Geophysical Research,* **64**:11 (November 1959).

12. E.P. Ney, J.R. Winckler, and P.S. Frier, "Protons from the Sun on May 12, 1959," *Physical Review Letters,* 3:4 (1959).

13. B.T. Price, C.C. Horton, and K.T. Spinney, *Radiation Shielding,* London, Pergammon Press, 1957.

14. W.L. Brown and G.L. Pearson, "Proton Radiation Damage in Silicon Solar Cells—Cases 38139-7 and 38788," Company Memorandum, Murray Hill, N.J., Bell Telephone Labs, (June 30, 1960).

15. W.L. Brown and G.L. Pearson, "Electron Radiation Damage to Silicon Solar Cells—Cases 38139-7 and 38788," Company Memorandum, Murray Hill, N.J., Bell Telephone Laboratories, (April 6, 1960; contains Lofersky and Rappaport data).

16. R.G. Downing, "Electron Bombardment of Silicon Solar Cells," ARS Report No. 1294-60 (September 1960).

17. A set of preliminary data curves received from the International Rectifier Corporation, El Segundo, California, September, 1960.

18. H.J. Schaeffer, "Radiation Danger in Space," *Astronautics* (July 1960).

19. H.J. Schaeffer, a paper presented to the 2nd International Symposium on Physics and Medicine of the Atmosphere and Space (November 1958).

20. K.A. Anderson, "Prediction Aspects of Solar Proton Events," July 1960.

21. K.A. Anderson, "Prediction Aspects of Solar Cosmic Ray Events—Preliminary Version," October 4, 1960.

A FAMILY OF RADIOISOTOPE-FUELED AUXILIARY POWER SYSTEMS FOR LUNAR EXPLORATION

ROBERT J. WILSON*

Assistant Project Engineer, Nuclear Division, The Martin Company, Baltimore, Md.

Designs for a family of auxiliary power systems have been developed under the SNAP program sponsored by the Atomic Energy Commission.[1] In all cases thermal energy derived from the decay of a selected radioisotope is converted directly to electrical energy by static thermoelectric conversion systems.

After demonstration of the initial prototype unit (SNAP 3) at the White House on January 16, 1959, several missions were proposed as reasonable applications of this type of power source in space. However, the safety of such a device had to be conclusively demonstrated under any contingency affecting its integrity. Subsequent positive ground tests simulating every phase of a missile malfunction to which the unit might be subjected have been conducted with satisfactory results. The next step will be actual flight tests, in which generator components (but no radioactive fuel) will be ejected from ballistic missiles. By monitoring the re-entry of the components, it will be possible to determine the validity of heating rate assumptions and dispersal calculations. These tests should remove all technical reservations concerning the safety of these units in combination with a vehicle system.

It is the intent of this paper, however, to emphasize the current capability of nuclear power, rather than to discount the reservations which have existed.

All important space missions to date have relied upon chemical or solar sources for the production of electricity, but several specific missions can be defined in which radioisotope sources have significant advantages. These include:

1. Satellites operating on equatorial orbits where the continuous power demand exceeds the storage cycle capability
2. Satellites operating for considerable periods of time in the Van Allen radiation belts where solar cell output is significantly degraded

---
*Acknowledgement is made of the efforts of many fellow employees of the Nuclear Division of The Martin Company in the development of the data described herein.

---
[1]SNAP—Systems for Nuclear Auxiliary Power, AEC Contract AT(30-3)-217

3. Lunar explorations, such as NASA's Surveyor program, where
    storage requirements during the lunar night become excessive
4. Venus surveys where solar energy would be excluded by the
    atmosphere
5. Space probes away from the sun where the decrease in solar
    flux becomes significant

Auxiliary power systems capable of fulfilling the demands of these
missions with radioisotope sources will be described as follows:

Designation	Electrical Power (watts)	Operational Life	Weight (lb)	Isotope Source
SNAP 1A	125	1 year	200	Cerium 144
SNAP 3	2.8	90 days	4	Polonium 210
Satellite	14.5	5 to 10 years	10.4	Plutonium 238
Soft Lunar Landing	13	6 months	16	Curium 242
Hard Lunar Impact	13	2 months	6	Curium 242
Space Probe	100	6 months	77	Curium 242

The first two have already reached the hardware stage, while designs for the others are complete and component tests which will
further the development of the final units are under way.

SNAP 1A Generator

SNAP 1A consists, structurally, of a central fuel container supported by tubular truss members within two concentric shells. Load
members between the shells provide strong points for ground-handling
and flight-vehicle attachment.   A cylindrical block of Inconel X,
3¾ in. in diameter and 11 in. long, contains the radioisotope. Seven
thin-walled stainless steel tubes loaded with ceric oxide pellets
are inserted into holes drilled in the block. Threaded closure caps
are screwed into place and later fusion-welded to provide a sealed
container.   The loaded block may be transported separately in its
own shipping cask and installed, with portable ground-handling
equipment, through a port in the generator at the launch site. Physical relationships are illustrated in Figure 1.

Heat generated in the central source is radiated to the inner
shell of stainless steel. The inner shell serves as a hot junction
for the thermoelectric elements installed between the shells. Un-
converted thermal energy conducted through the elements and es-
caping through the insulation is radiated from the outer aluminum

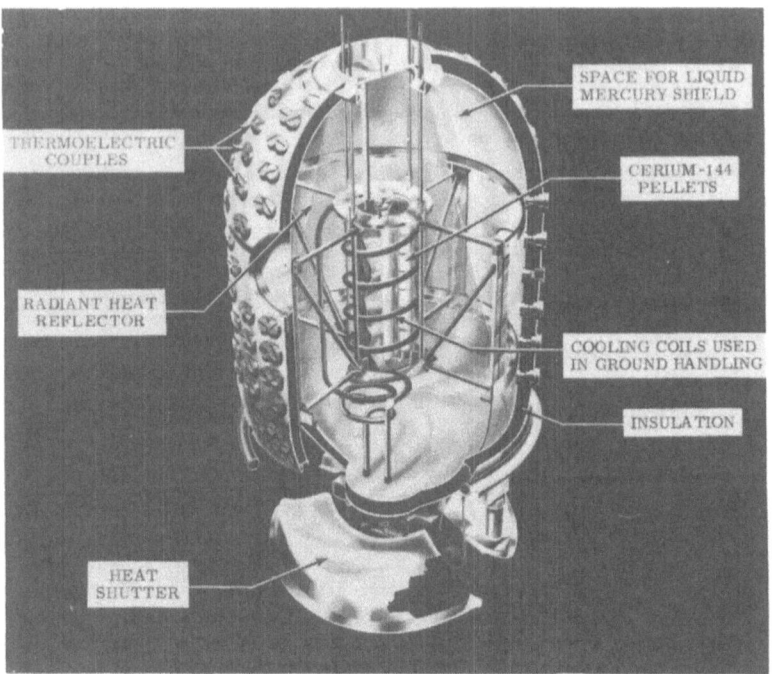

Figure 1. SNAP 1A.

shell. Excess energy, resulting from the excess fuel loaded initially to compensate for isotope decay, escapes from an exposed
area of the inner shell. Insulated shutters, actuated by a mercury
vapor bellows, control the amount of exposed area as required to
maintain a constant hot-junction temperature.

Approximately 6,000 thermal watts will be produced by an initial
load of 880,000 curies of cerium 144. Over a period of one year, the
output will decay to 2,700 w, the generator design condition. The
outer shell, 24 in. in diameter and 34 in. long, must possess sufficient area to radiate this energy to the space sink. At initial loading 3,300 w will be dumped from the exposed inner shell with the
shutters completely open. A fibrous insulation, Min-K-1301, is
used between the shells to reduce heat losses to a minimum.

The electrical features of SNAP 1A are as follows. Two hundred
seventy-seven thermoelectric couples are connected in series to
produce 125 w at 28 v. Each couple consists of two lead telluride

cylinders doped with either lead iodide, to form the *n* element, or sodium, to form the *p* element. An iron shoe connects the elements at the hot junction while copper caps are soldered to the cold junction ends. The elements are 0.375 in. in diameter, with lengths of 0.78 in. for the *n* and 0.66 in. for the *p*. Operating at a hot junction temperature of $1000^{\circ}F$ and a cold junction temperature of $335^{\circ}F$, these elements attain a thermoelectric efficiency of 6.75 per cent which, because of the thermal losses through the generator insulation and structural members, decreases to an over-all efficiency of 4.7 per cent. Significant design characteristics are summarized in Table 1.

TABLE I. SNAP IA DESIGN CHARACTERISTICS

Heat Source			
Radioisotope			Cerium 144
Quantity (curies)			880,000
Power output (electrical watts)			125
Operational life at constant power (days)			365
Voltage (volts)			28
Current (amperes)			4.46
External load (ohms)			6.28
Efficiency (per cent)			
Thermoelectric			6.75
Over-all			4.7
Thermal energy (watts)			
At initial loading			6000
At end of operational life			2700
Operating temperature ($^{\circ}F$)			
Hot junction			1050
Cold junction			335
Radiator surface (square feet)			
Low temperature			14.5
High temperature			3.5
Weight (pounds)			
Hg shield (removable)			4000
Flyaway			200
Thermoelements	*n element*		*p element*
Diameter (inches)	0.375		0.375
Length (inches)	0.78		0.66
Doping	Lead iodide		Sodium

Cerium 144 was selected as the radioisotope heat source for this generator. The basis for selection was its availability, being a fission product separated from reactor wastes at Oak Ridge National Laboratory. It decays through beta emission, with a half-life of 285 days, to praseodymium 144, which, in turn, decays through beta emission with a half-life of 17.5 minutes. The third element in the decay scheme, neodymium 144, decays with a half-life of 1.5×10^{15} years through emission of an alpha particle. Because of its long half-life, neodymium is considered to be stable. Although some of the thermal power originates from the initial cerium decay, the major energy source is derived from the energetic beta decay of praseodymium. Thus, in addition to the decay gammas, a large portion of the photon radiation results from the slowing down of these energetic beta particles *(bremsstrahlung).*

In reducing the radiation level to a reasonable amount for ground-handling purposes, sufficient volume exists between the heat source and the inner shell to contain approximately 4,000 pounds of mercury. Because of the source and shell geometry, shield thickness and dose rates vary from point to point around the unit. With the mercury in place, radiation levels 1 m from the generator surface vary from 2 to 95 mr per hour.

Two SNAP 1A units have been constructed for test purposes, using an electrical heater in lieu of a radioisotope heat source. A third unit, currently under construction, will be fueled and subsequently ground tested.

SNAP 3 Generator

This device was the first to produce a practical amount of electrical power by direct conversion of the heat resulting from radioisotope decay. It was demonstrated at the White House on January 16, 1959. Its characteristics are tabulated as follows:

Physical dimensions (inches)	4.75, diameter; 5.5, height
Weight (pounds)	5
Source (curies)	1495, Po 210
Half life (days)	138
Thermal power (watts)	48
Electrical power (watts)	2.4
Efficiency (per cent)	5

Thermoelectric material	Doped lead telluride
Hot-junction temperature ($^{\circ}$F)	720
Cold-junction temperature ($^{\circ}$F) (outer skin)	175
Dose rate at surface (mr/hr)	350
Dose rate at 1 m (mr/hr)	3.5

Polonium 210, with a half-life of 138 days, was selected as the radioisotope heat source for this unit on the bases of ready availability and high-power density. Its thermal energy results almost entirely from the emission of a 5.3-mev alpha particle. A low-gamma emission coincident with a 4.5-mev alpha particle occurs in 0.0012 per cent of the disintegrations. Mound Laboratory has provided the polonium for all generators fueled to date. Double stainless steel capsules are inserted in a heavy-walled heat source cylinder. This cylinder is then closed with a threaded plug and sealed by fusion welding. In this unit the thermoelements support the central source.

The Minnesota Mining and Manufacturing Company designed and fabricated the thermal-to-electrical conversion section. The lead telluride elements, approximately 1 in. long, have diameters of 0.210 in. (n-type) and 0.225 in. (p-type). These are paired to form 27 couples which are mounted in six vertical rows with a radial arrangement of nine elements per row. Each element is loaded axially against a hot shoe by a 2-lb force provided by a spring and adjustment screw located in an aluminum cold-junction ring. The hot shoe is iron which has been flame-sprayed with aluminum oxide. The shoe makes the electrical connection between element pairs at the hot junction, while the oxide coating prevents current flow through the supporting hot-junction cylinder. The couples are connected in series by copper wires at the cold junction and insulated electrically from the cold-junction ring. Insulation is provided by an oxide coating (Martin Hardcoat) on the aluminum bearing caps located between the element end cap and the pressure spring. The hot-junction cylinder supports the elements through the hot shoes and contains a tapered internal hole to receive the fuel capsule previously described. Mica sleeves are placed on the hot junctions of the elements to provide additional protection and insulation.

The remaining void inside the generator shell is filled with Johns-Manville Min-K-1301 insulation. This material is formed into

solid pieces in the hemispherical ends, and it is in powdered form
in the thermoelectric area. This internal arrangement is illustrated
by the cutaway artist's concept in Figure 2. After the first unit, a
spun copper shell was used instead of the brass case illustrated.
Other improvements in construction details decreased the weight to
4 lb.

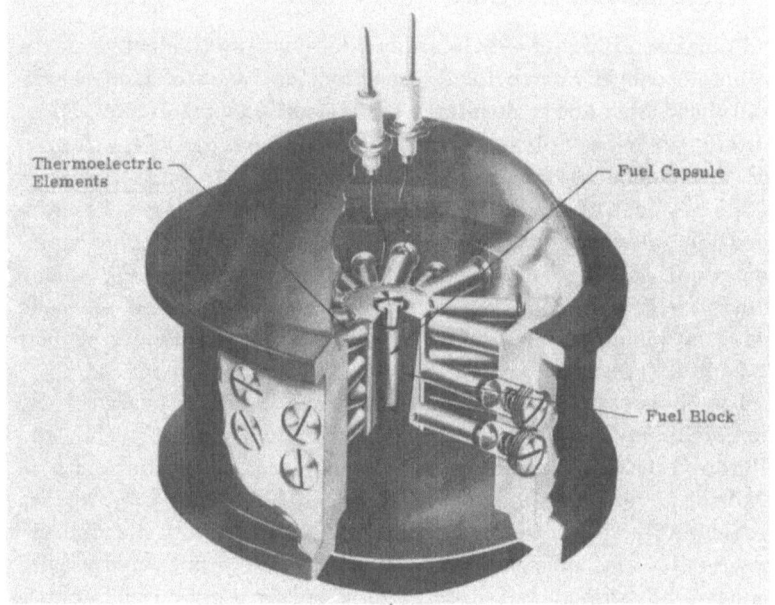

Figure 2. Low power radioisotope thermoelectric generator.

Three additional units have been fueled with various amounts of
polonium, ranging upward to 2,320 curies. A maximum output of 4 w
at 4 v has been obtained with an efficiency of 5.75 per cent. The
last two units were subjected to over 2,000 hours of severe environmental testing prior to being fueled. Their operation has been
satisfactory to date.

It may be noted that the Abstract lists the output of SNAP 3 to
be 2.8 w, versus the 4 w given above. This is the result of two
effects. First, in a space environment the isotope load must be
reduced slightly to avoid overheating. Since radiation is the only
method of heat transfer from the cold junction, its temperature in-

creases slightly over that in air, causing a corresponding increase in hot-junction temperature if the same thermal input is provided.

Secondly, the electrical output would fall off to approximately 1.8 w over a 90-day period without some method of power-flattening. A method of obtaining a constant output has been devised through tests in Martin laboratories. Conductivity of the Min-K insulation decreases rapidly at pressures less than 1 atm. Thus, by introducing a conductive gas mixture of helium and hydrogen into the shell, the conductivity may be increased initially, and then decreased by allowing the gas to leak to space. This conductivity change may be controlled, by proper selection of a fixed orifice, so as to be inversely proportional to thermal energy decrease. Initially, with a gas charge of 1 atm, heat losses through the insulation are high, and generator output is decreased to approximately 2.8 w. As this gas charge leaks out, generator efficiency improves so that a constant output is produced for a 90-day period.

Safety Tests

Although performance and environmental tests have defined the behavior of these generators under normal flight conditions, an essential adjunct has been the determination of the safety of these units in combination with a vehicle system. For logical planning of a test program to simulate the critical situations of any mission, the vehicle trajectory was divided into three phases: launch, ascent, and orbital injection or re-entry. Each phase was then analyzed to establish test criteria encompassing the greatest hazards to the integrity of the unit.

Critical situations existing on the launch pad arise from the explosive potential of a completely fueled missile. This potential is not fully realized unless a mass spill occurs through tank rupture, due to fall-back after lift-off, inadvertent actuation of the destruct system, or structural failure. The general natures of the phenomena occurring depend on the type of fuel used. Reports published by Space Technology Laboratories, Inc.[1], and personal contacts with their personnel were very helpful in defining these phenomena, which, by their very nature, are difficult to analyze. For test purposes the over-all reaction was divided into two basic effects: shock overpressure and high temperature.

Ignition of a mixture of liquid oxygen and RP-1 will produce a significant shock wave. Although the impulse characteristics are

slightly different, a TNT equivalent was computed with the assistance of Dr. W. Baker of the Ballistics Research Laboratory at Aberdeen Proving Ground[2]. Specimen size, quantity of explosive, and
distance from the center of explosion to specimens were scaled one-
third size to reduce explosive requirements. Generator and capsule
specimens were arranged at a 12-ft distance around 1,650 pounds of
TNT. The measured shock overpressures created by its detonation
exceeded those measured in actual missile failures by 125 per cent.
Effects on the isotope capsules were negligible although the external generator shells were destroyed.

Mixing of hypergolic propellants produces extreme temperatures
around 6000° F without significant shock. This situation was
created with 165 gal of aniline and 244 gal of nitric acid. These
liquids were supported on the top platform of a wooden tower under
which 20,000 pounds of scrap metal were stacked to simulate missile structural materials. Explosive charges were used to rupture
the liquid containers simultaneously. Temperatures ranged down-
ward from 6000°F, decreasing to 2000° to 1500° F after 20 minutes.
Generator capsule specimens, all full-sized, were suspended in the
fire zone. Capsule integrity was maintained in all specimens.

The second critical phase occurs during ascent of the vehicle.
Propulsion failure or command destruction will result in impact of
the generator on the earth's surface at various velocities at various
distances down range. For test purposes a terminal velocity of 500
ft/sec for the isotope capsule and 350 ft/sec for the generator assembly were assumed to be the maximum attainable. Generator
specimens were impacted against targets ranging in hardness from
granite to water. Tests were conducted on the ballistic track at
Aberdeen Proving Ground, where FFAR rocket motors accelerated
the specimens to the desired velocities. Although considerable
abrasion and deformation occurred in some capsule specimens, pressure tests proved that capsule integrity had been maintained.

After impact, the isotope container is exposed to a variety of
corrosive media existing on the earth's surface. The most common
of these are considered to be an oxidizing atmosphere, salt water,
and salt spray. Heated specimens and unheated control samples of
capsule materials were exposed to these environments. Testing was
halted after 720 hours since the Haynes 25 and Inconel X specimens
developed an oxide coating which, once formed, appeared to inhibit
further oxidation.

Normal operation of the vehicle system will place the payload and generator in orbit, but with the current methods predicted, orbital lifetimes are difficult to achieve or control.   Therefore, re-entry must be assumed at any time during the planned operational life of the generator.   In addition, malfunction of the final stage could result in velocities less than adequate for orbit or misaligned thrust vectors, causing immediate re-entry.   Detailed studies led the SNAP Hazards Committee and the Aerospace Nuclear Safety Committee to recommend design of the unit for aerodynamic burn-up and consequent dispersal of the radioisotope at the earliest time, highest altitude, and smallest particle size attainable.   The temperature, velocity and heat input calculated for typical trajectories were simulated in plasma arc facilities in the Martin Laboratory and at the General Electric Aeroscience Laboratory.   The destruction characteristics of the generator and isotope capsule were determined, as well as the resulting fuel particle sizes.   These data have been correlated with the vehicle trajectories to determine dispersion altitudes and locations.   Upon release, these particles behave like weapons fallout, although of a much lower order of magnitude.   In the case of SNAP 1A approximately 5 millicuries per square mile would be deposited, in comparison to the 2,000 millicuries per square mile already present.   SNAP 3 results would be virtually undetectable.

These tests have conclusively demonstrated the safety of these SNAP systems within the capability of ground equipment to create a re-entry environment.   The next step will be actual flight tests in which all generator components except the radioactive material will be ejected from ballistic missiles.   Monitoring of component re-entry will determine the validity of heating rate assumptions and dispersal calculations.   The extensive test program to date and that planned in the near future are indicative of the measures sponsored by the Atomic Energy Commission to dispel the initial reservations associated with the use of these devices in space.

Satellite Generator

The configuration of this unit is, in general, similar to that of SNAP 3.   This approach was adopted in order to reduce the development time required and to increase reliability.   An operational life of 5 to 10 years in orbit was the design objective (see Figure 3).

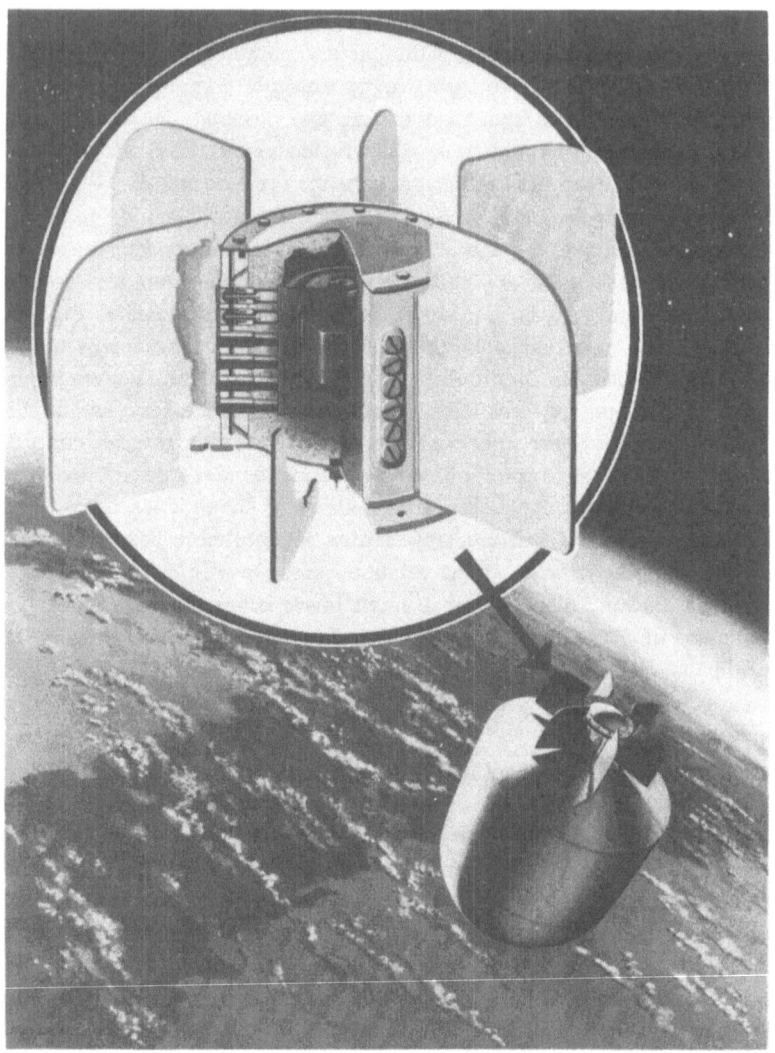

Figure 3. Satellite generator.

Plutonium 238, in carbide form, was selected as the radioisotope heat source since its half-life of 86 years obviates the need for a heat dump or power-flattening mechanism. Because of the relatively lower power density (6.90 w/cc) of plutonium carbide, the central heat source (a right circular cylinder 2.8 in. in diameter) is large in comparison to the over-all generator size.

Conversion of the thermal energy will be accomplished at an efficiency of 6.54 per cent. Lead telluride elements arranged in 36 couples will produce 14.5 w at 3 v. A hot junction temperature of $900^{\circ}F$ was selected to assure satisfactory element operation over the operational lifetime. In order to maintain an efficient temperature differential, the cold junction temperature was lowered to $235^{\circ}F$ through extended radiator surfaces. Six fins, extending 3½ in. from the outer shell surface, increase the effective radiation area.

Heat losses through the insulation at operating temperatures were calculated to be 40 w, placing the total energy required at 262 w. Combining this thermal efficiency of 85 per cent with the thermo-electric efficiency gives an over-all efficiency of 5.5 per cent.

The estimated weight of this unit is 10.4 lb with a specific power of 1.4 w/lb.

This unit is particularly attractive where extended exposure to Van Allen radiation occurs. The performance of solar cells degenerates significantly when the integrated dose exceeds 5×10^{13} electrons per square centimeter [3].

Soft Lunar Landing Generator

Development of the generator was coordinated with the Jet Propulsion Laboratory of the California Institute of Technology, so that the unit would meet the general requirements of space probe packages, under concurrent development at CIT. The criteria for the conceptual design proved to be of the utmost significance in the work which ensued. These criteria were:

1. Electrical power — The generator was to deliver 13 w at 3 v direct current of continuous, unregulated electrical power for a period of at least 6 months, following launching into space.

2. Weight — Not to exceed 15 lb.

3. Environment — The unit was to operate in normal fashion in an undefined space trajectory (for example, a circumlunar probe), following subjection to missile-induced conditions typical of the Vega launch vehicle. Development of the Vega vehicle was canceled during the course of the program; these conditions were then held to be illustrative only.

4. External radiation — Because of the presence of radiation-sensitive equipment in the JPL payloads, external radiation emana-

ting from the generator was specified not to exceed 7 gamma photons/cm^2/sec above 100 kv at a point 10 cm from the surface. In the case of an anisotropic radiation field, the generator could be oriented with respect to the rest of the payload so that the minimum radiation level would be seen by the payload. No restriction was placed on neutron fluxes external to the generator during its operational life. For safe ground-handling, external radiation levels were to be reduced to a maximum of 60 mrem/hr at a distance of 1 m.

5. Fuel form — Curium 242 was specified as the radioisotope to be evaluated in the study. An extensive program aimed at developing this isotope as a heat source was being undertaken concurrently under Contract AT(30-3)-217. The chemical form or geometry to be employed was not specified, and was to be determined by the state of curium technology and the demands of the generator design. Its half-life of 163 days and specific thermal power of 122 w/g make curium 242 a suitable heat source for minimum-size generators having moderate operational lives of 3 to 6 months. Biological-shielding problems are minimized with Cm 242 because of the relatively low level of ionizing radiation emanating from the sealed source of the isotope.

The configuration of this unit resembles SNAP 1A, in that a central heat source radiates its energy to the hot junction and a movable thermal shutter controls hot-junction temperature. Two novel features were used in the design. First, the high-temperature heat-dump radiator forms an integral part of the heat source. This reduces radiator size to a minimum, since its temperature will be 1300 to 1500° F. Second, a block of tungsten shield material was also incorporated as part of the heat source. Generator characteristics are as listed in Table 2.

Recently, this design was used as a basis for proposals to the four contractors engaged in the NASA Surveyor program study. System power requirements would be supplied by one to four of these units, depending on contractor vehicle arrangements.

In all cases, the original power output of 13 w was not adequate. Initial investigations of methods to increase the power produced by the unit were based on the original thermal efficiency of 73.5 per cent. Results of these studies indicated that 19 w could be attained by increasing the thermoelement size and radiator area. Eight axial aluminum fins, 0.08 in. thick, extending 3 in. from the original radiator surface over its length of 7.5 in., provided adequate additional area.

TABLE 2 [4]. GENERATOR CHARACTERISTICS

Operational life (months)	6
Size	
Diameter (inches)	7.5
Height (inches)	8.38 (shutter closed)
Efficiency (per cent)	
Thermoelectric	6.51
Thermal	73.5
Over-all	4.8
Temperatures ($^\circ$F)	
Hot junction	1000
Cold junction	370 (in space)
Heat loss (watts)	36
Insulation thickness (inches)	1.25
Thermoelectric elements	
Output voltage (volts)	3
Number of pairs	30
Cross-sectional area (square inches)	
p-type	0.1300
n-type	0.1089
Doping	
p-type	1.0%
n-type	0.03% PbI
Length (inches)	0.75
Fuel	
Isotope	Curium 242
Purity (per cent)	45
Dilution (Au to Cm by weight)	5 to 1
Mass of isotope (grams)	6.3
Thermal power (watts)	
At encapsulation	752
At launch	582
At end of life	270
Estimated weight (pounds)	16.64

In most cases, radiation-sensitive instrument placement within the spacecraft made the internal gamma shield material ineffective. Therefore, this shield was removed in favor of individual shields at the instruments.

Subsequent studies also revealed that the thermal efficiency of the generator could be increased to 81.4 per cent through incorporation of the following design improvements, illustrated in Figure 4.

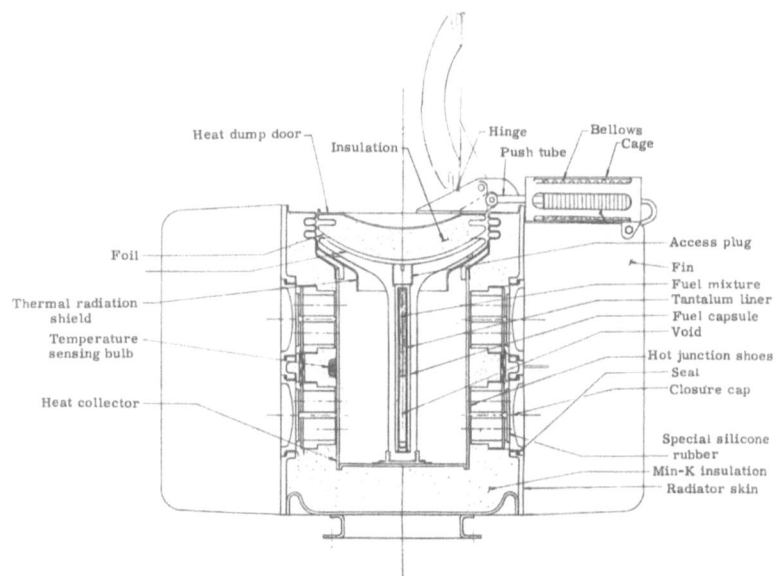

Figure 4. Soft lunar landing generator.

1. After removal of the tungsten shield material, the fuel capsule and heat source could be reduced in size.

2. A thermoelectric module concept would be used where four elements are bonded together to allow external installation without depending on a pressure contact for an electrical connection. Aside from the many advantages inherent in this concept from an assembly and fabrication standpoint, a much more rugged and reliable unit results.

The heat-dump door and its actuator would be revised to reduce the heat losses in the closed position. In addition, a mercury vapor system would be lighter than the original sodium-potassium mixture.

Thermoelectric element data for a 19-w output are given in Table 3. Assuming the radiator surface to have a solar absorptivity of 0.4 and an infrared emissivity of 0.94 gives solar and lunar thermal inputs of 32 and 50 w, respectively. An emissivity of 0.9 was assumed for the lunar surface with generator orientation normal to both radiation sources. With these inputs, cold-junction temperatures of 400 and 390° F were estimated for the lunar day and night, respectively. Therefore, generator operation will be relatively invariant between day and night conditions.

TABLE 3. THERMOELECTRIC ELEMENT DATA

Power to the load — 19 w	**Positive Element Data** Element length — 0.75 in.
Current — 6.33 amp	Cross section at hot junction — 0.247 sq in.
Open circuit voltage — 5.44 v	Diameter — 0.56 in. Shape factor (A/l)* — 0.33 in.
Number of couples — 31	Internal resistances (milliohms) Element — 4.73
Generator total resistance — 0.38 ohm	Contact — 1.68
Voltage across load — 3 v	**Negative Element Data** Element length — 0.75 in.
External load — 0.47 ohm	Cross section at hot junction — 0.19 sq in.
Temperature (°F)	Diameter — 0.497 in. Shape factor (A/l) — 0.258 in.
Hot junction — 1,000 Cold junction — 370	Internal resistances (milliohms) Element — 5.04 Contact — 0.917

* A = Cross-sectional area
l = Thermoelement length

From the data of Table 3 thermoelectric efficiency was calculated to be 5.35 per cent. However, the estimated cold-junction temperature exceeds that used by 30° F. Direct application of a correction factor for this decrease in temperature differential causes the thermoelectric efficiency to decrease to 5.10 per cent. Using this value in combination with a thermal efficiency of 81.4 results in an over-all generator efficiency of 4.15 per cent at the end of operational life, with the heat-dump door fully closed. A thermal input of 458 w will be required to produce 19 electrical watts under these conditions. According to the exponential decay law, the initial capsule loading at 130 days prior to end of operational life will produce 800 thermal watts. With a power density of 122 w/g, 6.6 g of curium 242 will be required.

As a result of the design improvements previously enumerated, generator weight was estimated to be 12.5 lb, with a corresponding power density of 1.5 w/lb.

Hard Lunar Impact Generator

The conceptual design of the lunar impact generator was to be based on the following specifications [5]:

1. A maximum weight of 18 lb.
2. A continuous output of 13 w at 3 v direct current for a period of 60 days after impact.

3. Ability to withstand the decelerative forces resulting from lunar impact at 500 ft/sec.

4. Maximum reliability and efficiency consistent with the other specifications.

5. Use of curium 242 as the radioisotope fuel.

6. Assurance of fuel capsule integrity under adverse conditions of launch vehicle abort;

7. Reduction of external radiation level to 60 mrem/hr at 1 m for ground-handling purposes, and to the following levels in operation for protection of radiation-sensitive equipment in the instrument capsule: On a circular area of 100 cm^2 located 5 m from the generator, radiation was not to exceed:

    a. 1 photon/cm^2-sec in the energy range 0.4 to 3.0 mev.

    b. 0.5 photon/cm^2-sec in the energy range 3.0 to 10.0 mev.

8. Loading of the fuel capsule occurs 60 days (maximum) prior to launch.

Since details of the launch vehicle; impact vehicle, and instrument capsule were not available during the course of this study, it was necessary to make some arbitrary assumptions concerning them so that the generator design would at least reflect representative conditions for launch, abort, or lunar impact.

The impact vehicle was assumed to consist of a retrorocket attached to a spherical instrument capsule roughly 3 ft in diameter, consisting primarily of a crushable structure to cushion the impact forces on the instruments. Since the crushable material might possess thermal insulation properties, the generator was assumed to be mounted on the exterior of the capsule, at a point farthest removed from the retrorocket. In a successful mission, the retrorocket case would probably impact first.   Uniform deceleration of the payload was assumed to occur over a distance of 1 ft, leading to a figure of about 4,000 earth g deceleration. To be conservative, the generator design was based on an impact angle varying to a maximum of 15 degrees from the axis of the vehicle. The design that evolved was symmetrical about a linear axis which coincided with the axis of the impact vehicle, and maximum stresses were taken to be compressive, along the axial direction.

The resulting generator configuration is shown in Figure 5. It consists of a fuel block in the shape of a rectangular parallelepiped, containing the curium fuel in four sealed canisters.   Two modular arrays of thermoelectric elements are mounted on opposite sides of

Figure 5. Hard lunar impact generator.

the fuel block, and the remaining internal volume of the generator is filled with solid thermal-electrical insulation. The rigid outer shell, in the general shape of an ellipsoid of revolution, conducts rejected heat to radiator surfaces and is attached to a shock-absorbing bellows structure which, in turn, is fastened to the impact vehicle. Simplicity of the design and the lack of moving parts are apparent.

Results of the design analysis are tabulated in Table 4. A specific power of 2.1 w/lb may be attained after 60 days of operation while specific power at launch would be 2.9 w/lb. Although this unit and the soft lunar landing unit described previously were designed for different missions, their basic differences are due to the thermoelectric material characteristics (lead telluride versus a higher temperature material). All the advantages of the impact design are due to the ability of its thermoelements to operate in a higher temperature region. The exception is the improved shock resistance due to superior physical properties.

TABLE 4 [5]. RESULTS OF LUNAR GENERATOR DESIGN ANALYSIS

End of Life Parameters in 250° F Ambient — Lunar Environment	
Efficiency (per cent)	
Thermoelectric	5.36
Thermal	95
Over-all	5.2
Temperature (°F)	
Hot junction	1400
Cold junction	458
Fuel centerline	1500
Heat loss (watts)	17
Voltage, end of life (volts)	3
Power, end of life (watts)	13
Number of elements	127
Weight (pounds)	6.2
60 Days Prior to End of Life (at launch)	
Maximum temperatures (°F)	
Centerline	1800
Hot junction	1660
Cold junction	476
Power (watts)	18

Performance characteristics as described above are entirely feasible through development efforts in the following areas:

1. Reproduction of current laboratory materials on a "production" scale.

2. Generator and couple fabrication techniques to utilize these materials with low internal circuit resistances.

3. Determination of the life and stability factors of recently discovered materials.

4. Improvement of the figure of merit and physical properties of current material combinations.

A generator utilizing a high-temperature material could operate more efficiently without a variable heat dump, provided the load cycle of the system is properly programmed. A constantly decreasing power output from the generator would increase voltage control difficulties but would make a more reliable system as the thermal shutter mechanism could be eliminated.

Space Probe Generator

This design exercise was conducted to establish the characteristics of a larger generator using curium 242 as its heat source. No specific application was assumed, and the integration of this unit with a vehicle or power system was not investigated. A power output of 100 w for an operational life of 6 months was arbitrarily chosen.

Several alternate configurations were studied with the results indicating that an arrangement similar to that of SNAP 1A offered the highest specific power. The characteristics of this design are listed in Table 5. With a weight of 77 lb a specific power of 1.3 w/lb may be attained. Although no significant technological advances were used to develop this design, its specific power represents a significant improvement over that of SNAP 1A (0.625 w/lb). Two factors contribute to this improvement:

1. The higher specific power of curium 242 versus cerium 144.

2. A reduction in the mass of structural members since a mercury biological shield need not be supported.

TABLE 5 [6]. CHARACTERISTICS OF THE 100-w Cm 242 GENERATOR

Outer shell (cold junction) — aluminum	
Thickness (inches)	0.040
Height (inches)	21.5
Width, diameter (inches)	19.5
Inner shell (hot junction) — stainless steel	
Thickness (inches)	0.050
Height (inches)	19.5
Diameter (inches)	17.5
Thermoelectrics — pairs of PbTe elements	169
Length (inches)	0.3937
Diameter	
p couple (inches)	0.500
n couple (inches)	0.455
Doping	
p couple (per cent Na)	1.0
n couple (per cent PbI_2)	0.03
Heat source	
Height (inches)	4.5
Effective diameter (inches)	4.5
Fuel tubes, number	5
Diameter (inches)	0.956
Height (inches)	3
Height of fuel pellet (inches)	2.4 (20% void space)
Thermal power at encapsulation (watts)	7500
Fuel mixture, by weight	20:1 Ni-Cm mixture
Encapsulating material	Hastelloy C
Thermal shutter	
Area (square feet)	1.72·
Insulation — Min-K-1301	
Thermal conductivity (Btu-in./hr-ft^2-$^\circ$F)	
(vacuum)	0.1
Thickness (inches)	0.9
Density (lb/ft^3)	20
Temperature ($^\circ$F)	
Hot junction	1000
Cold junction	490
Heat source at launch	1800
Heat source at end of life	1470
Weight (pounds)	77
Efficiency (per cent)	
Thermoelectric	5.20
Over-all	4.95

REFERENCES

1. W.M. Smalley and D. E.° Anderson, "The Explosive Potential of Liquid Oxygen and RP-1 Missiles, GM-TR-59-0000-00579.
2. Ballistic Research Laboratory Report No. 1092, Aberdeen Proving Ground.
3. R. L. Sohn, "Review of Solar Cell Power Systems," presented at the AIChE Symposium on Unconventional Sources of Electrical Energy, New York, October 20, 1960.
4. J.L. Bloom, "13-Watt Curium-Fueled Thermoelectric Generator for Six-Month Space Missions," The Martin Company, MND-P-2372 (July 1960).
5. J.L. Bloom, "13-Watt Curium-Fueled Thermoelectric Generator for Hard Lunar Impact Mission," The Martin Company, MND-P-2374 (July 1960).
6. J.L. Bloom and J.B. Weddell, "100-Watt Curium-242 Fueled Thermoelectric Generator—Conceptual Design," The Martin Company, MND-P-2342 (May 1960).

EXTENDING THE RANGE OF RADAR-BEACON TRACKING FOR LUNAR PROBES

NORMAN S. GREENBERG

Microwave Equipment Laboratory, ACF Electronics Division, ACF Industries, Incorporated, Paramus, New Jersey

Present-day radar-beacon tracking equipment is limited to a range of less than 2,000 miles. To extend the range of the radar and beacon systems, increased transmitter power output, receiver sensitivity, and antenna gain are required. A review of the range equation shows the range capabilities of an existing radar and beacon, and allows prediction of the requirements of a system capable of tracking a vehicle to the vicinity of the moon and beyond.

Work has already begun to provide higher-power transmitters for both the radar and beacon. These are described.

Since the sensitivity of the receiver can be increased substantially by decreasing the bandwidth requirement, a nomograph is provided which gives the graphical relationship between sensitivity, noise figure, and bandwidth. The practical limitations of bandwidth as a function of local oscillator stability is discussed.

Specific attention is given to the possible modifications of instrumentation radar AN/FPS-16 to allow tracking a beacon to extreme ranges. The proposed beacon will have a peak power output of 40 kw and a triggering sensitivity of −75 dbm. Special components required for the beacon are described.

When using a single radar for tracking the beacon, range accuracy of ±25 yd can be measured at lunar distances. The system is capable of measuring angles to within 0.1 mrad.

Variations in the modes of radar-beacon operation will provide improved accuracy of tracking. A method is described which will allow locating a point on the surface of the moon to an accuracy of 5,000 yd.

Radar beacons are used primarily to increase the range of tracking radars which provide missile and satellite trajectory information. In most current applications this tracking information allows the range safety officer to be informed of the missile's path in order to take appropriate action should the missile go off course.

In other applications the tracking data are used to determine the time for remotely activating the engine ignition or fuel cutoff system to control the missile's trajectory. The radar-beacon link has also been used to provide command functions for initiating maneuvers of the missile or satellite.

In the Atlas/Agena/Ranger lunar probe, to be carried out by the NASA, radar-beacon data will be used for determining the trajectory of the first two stages. After first-stage separation, the Agena's engine is powered for a short interval and then automatically turned off. The vehicle continues to coast in what is called a "parking orbit." At a later time there is a second ignition of the Agena's engine in order to set the vehicle on a proper lunar inclination. During this operation, radar-beacon data are fed to a computer, where computations are made to determine that the missile is on course.

All of these applications of radar beacons require a conservatively rated maximum operation range of less than 2,000 miles. It will be shown that it is feasible to design a radar-beacon system having parameters capable of tracking a vehicle to the moon. Each of the parameters effecting the range capability of the system are analyzed.

Beacon Range

The beacon-range equation and a nomograph for quickly performing the calculation are given in the Appendix.

Typical values for an existing system operating at C-band frequencies (5,400 to 5,900 mc) are shown in Tables 1 and 2. The radar considered is Instrumentation Radar AN/FPS-16. The beacon is the Type 149C developed for the Mercury program (see Figure 1).

A maximum range of 1,700 statute miles, as determined by the response link, is possible for a radar receiver with an acceptable minimum signal-to-noise ratio of 15 db. Ranges up to 6,000 miles can be achieved, but accuracy is diminished due to the lower signal-to-noise ratio. Instrumentation radar AN/FPS-16 normally requires a signal-to-noise ratio of 17 to 20 db to track a radar beacon accurately. With this ratio the radar is capable of measuring range to within 5 yd, and angle to within 0.1 mrad.

To extend the range of the system to 250,000 miles — that is, to the surface of the moon — at least 32 db more is required in the interrogation link and 43 db in the response link. This improvement can be accomplished by providing greater transmitter power, more receiver sensitivity, and increased antenna gain for both the radar

TABLE 1. RANGE CAPABILITIES OF C BAND INTERROGATION LINK

	Existing Equipment	Proposed Equipment
Radar transmitter power	10^6 watts pk	3×10^6 watts pk
Radar antenna gain	44 db	52 db
Beacon receiver sensitivity	−65 dbm	−75 dbm
Beacon antenna gain	0 db	20 db
System losses	12 db	12 db
Tracking range with 100% triggering of beacon transmitter	6,000 statute miles	850,000 statute miles
Reserve system gain for lunar range of 250,000 statute miles	−32 db	+ 11 db

TABLE 2. RANGE CAPABILITIES OF C BAND RESPONSE LINK

	Existing Equipment	Proposed Equipment
Beacon transmitter power	10^3 watts pk	40×10^3 watts pk
Beacon antenna gain	0 db	20 db
Radar Receiver sensitivity	−99 dbm	−109 dbm
Radar antenna gain	44 db	52 db
System losses	12 db	12 db
Tracking range with 15 db s/n allowed for radar receiver	1,700 statute miles	850,000 statute miles
Reserve system gain for lunar range of 250,000 statute miles	−43 db	+11 db

and beacon. The proposed values of radar and beacon parameters shown in Tables 1 and 2 will allow tracking to the moon.

In the interrogation link the proposed system has a reserve gain of +11 db. Since the beacon receiver sensitivity is actually a measure of the signal required for 100 per cent triggering of the beacon transmitter, this reserve gain is more than sufficient for reliable operation.

Figure 1. Radar beacon, type 149C.

The radar receiver sensitivity of −109 dbm is specified as the tangential sensitivity, which is equivalent to a signal-to-noise ratio of unity. Although this signal level can be easily distinguished on an A-scope by an operator, it is usually not sufficient for controlling the automatic range circuits and antenna-positioning servos. Accurate tracking is assured by specifying a signal-to-noise ratio of +15 db.

Transmitter Power

The radar transmitter currently used has a peak-power output of 1 megawatt with a maximum duty cycle of 0.001. It is proposed that the radar be equipped with an existing klystron transmitter having a peak power of 3 megawatts and a duty cycle of 0.002 − a design that has been proven in operation at White Sands Proving Grounds. The radar should be capable of providing a pulse width up to 5 μsec. The bandwidth of the beacon receiver can then be made more narrow and thus provide greater sensitivity for a given noise figure.

Existing beacon transmitters deliver approximately 1 kw peak power. At C-band frequencies either a magnetron or a triode cavity oscillator is used. A beacon magnetron with a peak power of 40 kw can add 16 db to the response link. Such a tube is under development by the Army Signal Corps. For characteristics of this tube, see Table 3.

TABLE 3. CHARACTERISTICS OF HIGH-POWER BEACON MAGNETRON

Frequency range	5,625 to 5,675 mc
Peak power output	40 kw
Duty cycle	0.0003 maximum
Pulse width	10 μsec maximum
Anode efficiency	35%
Vibration	15 g up to 2,000 cps
Weight	5 lb maximum

Receiver Characteristics

The sensitivity of the radar and beacon receivers is dependent upon the noise figure of the front end, the receiver bandwidth, the conversion loss of the mixer, and the insertion loss of the preselector. Sensitivity is normally measured in minus dbm (decibels below 1 mw). Figure 2 shows the relationships for tangential sensitivity as a function of bandwidth for various noise figures. The characteristics of the existing and proposed receivers are shown. Figure 3 shows the noise figure that can be achieved for various types of receivers.

The AN/FPS-16 radar has a noise figure of less than 11 db and a bandwidth of 2 mc to provide a tangential sensitivity of −99 dbm. As in the case of AN/FPQ-6 radar, the proposed radar receiver will use a parametric amplifier having a noise figure of 6 db. With a bandwidth of 850 kc this corresponds to a tangential sensitivity of −109 dbm. In time it is expected that a MASER will be designed for the front end of the radar receiver which will provide a noise

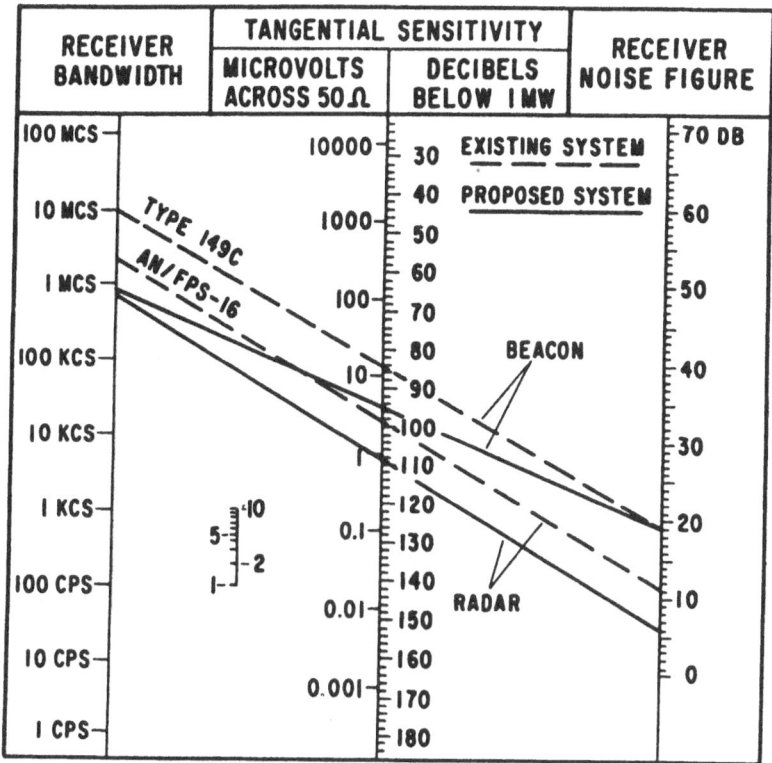

Figure 2. Nomograph — receiver sensitivity as a function of bandwidth and noise figure.

figure of 1 db, with a corresponding improvement in sensitivity of 5 db.

A typical C-band beacon uses a superheterodyne receiver having a tangential sensitivity of −85 dbm and a bandwidth of 10 mc. The noise figure is 19 db. Due to noise spikes produced by the power supply and other extraneous pickup in the beacon, the beacon is normally adjusted to trigger with a signal 15 to 20 db greater than the tangential signal level. Type 149C beacon is normally set to have a triggering sensitivity of −69 dbm which implies no countdown or extraneous triggering of the transmitter.

The proposed beacon receiver will require a tangential sensitivity of −95 dbm to secure a triggering sensitivity of −75 dbm with the bandwidth reduced to 850 kc. The equivalent noise figure of 19 db

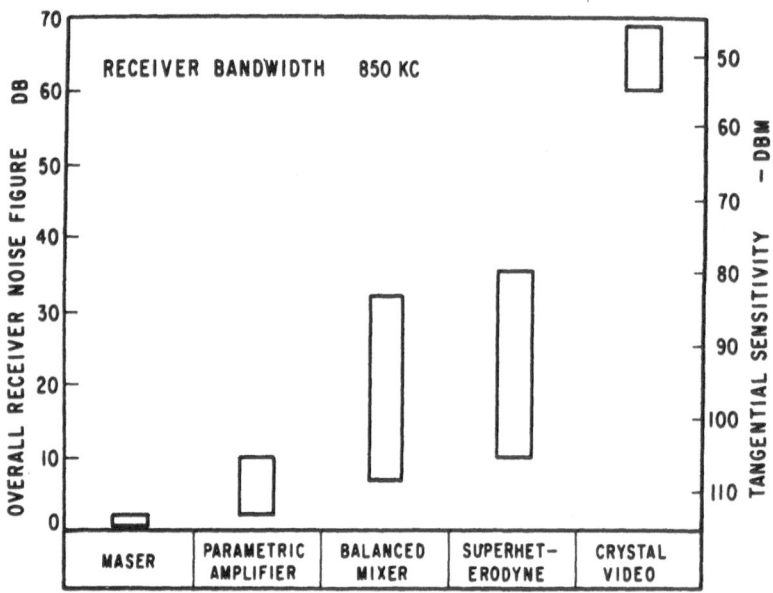

Figure 3. Over-all noise figures and tangential sensitivity for typical receivers.

can be assured with a balanced mixer. Reduction of the power supply noise and other interference within the beacon will allow operation with a signal-to-noise ratio of 20 db.

Alternate methods for obtaining the required sensitivity for tracking to the moon could use a parametric amplifier or MASER but consideration must be given to space and power consumption. These devices will become more applicable for space probes extending considerably beyond the range of the moon.

In current designs a bandwidth of 10 mc is necessary because of the narrow pulse widths used and the limited frequency stability of the radar transmitter and beacon local oscillator.

A radar transmitter having a stability of ±0.0001 per cent, which is equivalent to better than ±60 kc at C-band, can be designed. It will use a crystal oscillator operating at relatively low frequency. Frequency multiplication will provide frequencies in the range from 5,400 to 5,900 mc. The power output stage, which is already designed, is a klystron amplifier capable of providing up to 30 db gain and 3 megawatts peak power.

A more stable local oscillator is required for a narrow band receiver. The typical triode cavity local oscillator has a long term stability of ±2.0 mc under normal operating conditions. With the advent of the varactor diode it is possible to design a compact frequency multiplier which is stabilized by a crystal controlled oscillator operating at a VHF frequency. The long-term frequency stability of 0.005 per cent can be achieved, and this is equivalent to a local oscillator stability of ±300 kc. Since the optimum r-f bandwidth for a 5 μsec pulse is 240 kc, the receiver in the beacon should have a minimum total bandwidth of 840 kc.

Antenna Gain

Instrumentation radar AN/FPS-16 has a parabolic antenna 12 ft in diameter, which provides a gain of 44 db (see Figure 4). The gain of this type of antenna is given by the expression:

$$G = \frac{4\pi A f}{\lambda^2}$$

where A is the cross-sectional area, λ is the wavelength, and f is the dimensionless factor of approximately 0.6 for a parabolic antenna.

Figure 4. Antenna, instrumentation radar AN/FPS-16.

The gain of a C-band antenna as a function of the diameter of the parabolic reflector is shown in Figure 5. Plans have been proposed for the design of radar AN/FPQ-6, which will have a 30 ft dish and will provide an antenna gain of 52 db. This larger antenna will increase the range of the radar by a factor of 2.5.

The proposed long-range system requires a beacon antenna having a gain of 20 db. To accomplish this, an aperture area of less than 1 ft^2 is required.

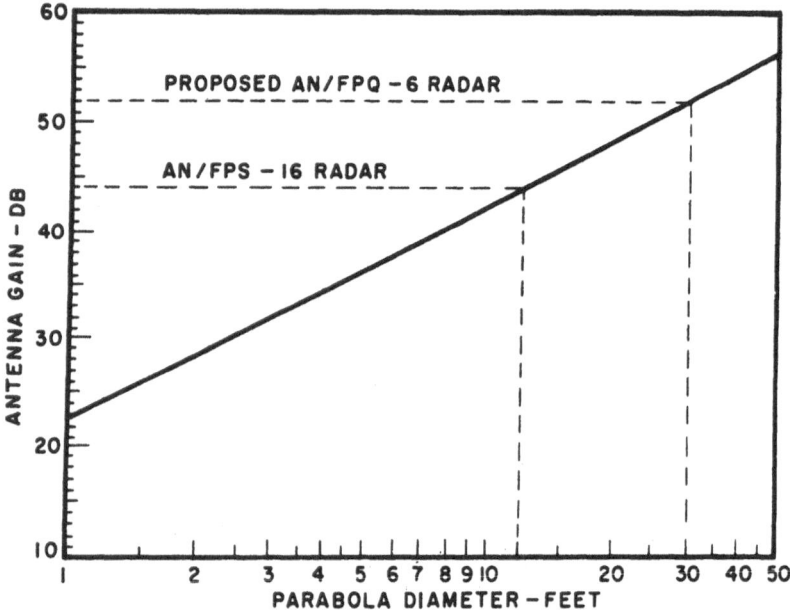

Figure 5. C-band parabolic antenna gain vs. diameter.

Angle Measurement

Radar AN/FPS-16 is capable of measuring azimuth and zenith angles to an accuracy of 0.2 mrad.[1] In the proposed system the repetition rate during long-distance tracking will not exceed 30 cps. Because of the extremely small value of $d\theta/dt$ of the missile, a relatively low repetition rate is satisfactory. The bandwidth of the radar's elevation and azimuth servos can be adjusted to approxi-

---

[1]See David K. Barton, "Accuracy of a Monopulse Radar," presented at the Third National Convention on Military Electronics, IRE, Washington, D.C., June 29, 1959.

mately 0.25 cps, and by smoothing techniques it will be possible to maintain an error of less than 0.1 mrad. The smoothing technique also decreases the required signal-to-noise ratio of the radar receiver and thus makes the proposed system performance practical.

Description of Beacon

The proposed long-range beacon will utilize special components and techniques. Many of these are being evolved under the sponsorship of the Army Signal Corps.

The proposed beacon will differ from the present-day standard unit in the following ways:

1. The transmitter will use a high-power magnetron having a peak power output of 40 kw. Both the magnetron and suitable modulator are being developed.

2. A special duplexer will be required which can provide the additional 16 db isolation for the receiver when the transmitter is triggered. A ferrite circulator is under development for this purpose.

3. The receiver bandwidth will be less than 1 mc wide, and a special high-stability local oscillator utilizing a crystal controlled oscillator and varactor multiplier will be used.

4. The front end of the receiver will have a balanced mixer to provide an improved noise figure.

Aside from bandwidth considerations, the i-f amplifier, and video amplifier will be designed similar to current practices, utilizing transistors throughout.

The equipment will be powered from a 28-v source and will require approximately 150 w for its operation. About 40 lb in weight, it will be capable of operating reliably under the environmental conditions of space flight.

Modes of Operation

The radar beacon tracking system can be used in three ways— normal angle- and range-tracking with one or more radars; angle-tracking using two or more radars; or range-tracking using two or more radars. Table 4 shows the accuracies to be expected with the proposed system under each mode of operation. The slant range is assumed to be 250,000 miles, and the accuracy for measuring slant range and perpendicular distance are given. This perpendicular dis-

TABLE 4. THEORETICAL CAPABILITY OF THE PROPOSED
RADAR-BEACON SYSTEM FOR LOCATING A POINT
AT LUNAR DISTANCES

Mode of Operation	Accuracy of Measuring slant range to moon's surface	Accuracy of Locating a point on moon's surface
Normal radar tracking with a single radar	25 yd	44,000 yd (25 miles)
Triangulation with angle data from two radars separated by 2,500 miles	4,400,000 yd (2,500 miles)	44,000 yd (25 miles)
Triangulation with range data from two radars separated by 2,500 miles	25 yd	5,000 yd (2.8 miles)

tance is equivalent to the error in measuring distance on the surface of the moon.

Normal beacon tracking of range and angle requires only a single radar. The range accuracy is a constant regardless of range, provided signal-to-noise ratio is sufficient. In the proposed system, having a bandwidth of 850 kc, the accuracy for measuring slant range will be 25 yd. The absolute accuracy of the perpendicular component decreases as range increases. For the proposed system, which measures angle to within 0.1 mrad, this error amounts to 25 miles (44,000 yd) on the moon surface.

The beacon may be programmed for theodolite-type tracking. In this mode of operation the transmitter is free-running and can be tracked by two or more radars located along known base lines. By means of triangulation, complete tracking data can be compiled. If a base line of 2,500 miles is assumed, the accuracy for measuring slant range is 2,500 miles. The accuracy of the perpendicular component is approximately 25 miles. This mode of operation would eliminate the need for the high-powered radar transmitter and extremely sensitive beacon receiver.

The most accurate method for tracking extreme ranges uses range data derived from two or more radars located along known base lines.

With a slant-range accuracy of 25 yd for each of the radars, the perpendicular distance can be calculated to within 5,000 yd on the moon's surface. This precision would locate, from the earth, the smallest identified crater on the moon.

Conclusion

Radar-beacon tracking at lunar distances can be accomplished by modifying instrumentation radar AN/FPS-16 or proposed radar AN/FPQ-6, and developing a radar beacon having a more sensitive receiver and increased power output. A directional antenna is required for the beacon. A discussion of the parameters required for a long-range tracking system shows the feasibility of such a system. Work is already in progress for improving the characteristics of the radar. Based upon the state of development of the special components required, a prototype beacon could be produced within a year.

APPENDIX: RADAR BEACON RANGE NOMOGRAPH[2]

The following equation is used to calculate range for radar-beacon tracking. Figure 6 is a nomograph which quickly solves this equation. The equation states that:

$$R = \frac{\lambda}{4\pi} \left(\frac{P_T \; G_T \; G_R}{P_R} \right)^{1/2}$$

where P_T and P_R are the transmitter and receiver power, G_T and G_R are transmitting and receiving antenna gains and λ is the wavelength of radiation used. The interrogation link (radar to beacon) and the response link (beacon to radar) are solved individually.

Range Example

Find the range when given an interrogation link with

Radar peak transmitter power. . . . . 1 megawatt
Radar antenna gain . . . . +44 db
Radar r-f losses . . . . −6 db
Beacon receiver sensitivity. . . . . −65 dbm
Beacon antenna gain . . . .   0 db
Beacon r-f losses . . . . −6 db
Frequency. 5,600 mc

---

[2] Adapted from article appearing in *Electronics* (A McGraw-Hill Publication), Vol. 32, No. 36 (September 4, 1959) p. 60.

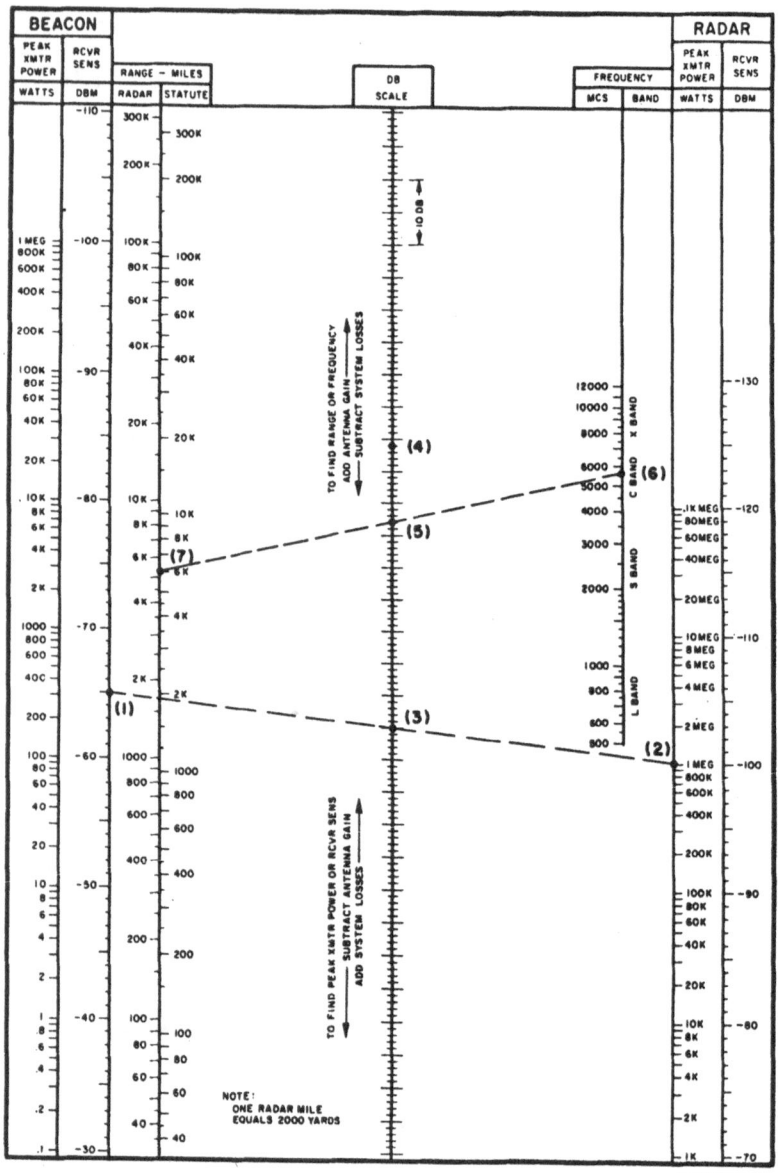

Figure 6. Radar-beacon range nomograph.

Draw a straight line between −65 dbm on *beacon recr sens* (1) and 1 megawatt on *radar peak xmtr power* (2). From the point where this line intersects the *db scale* (3), add, on this scale, the radar and beacon antenna gains totaling 44 db (4), less the radar and beacon r-f losses of 12 db (5). Extend a straight line from 5,600 mc on *frequency* (6) through (5) on *db scale* to *range* and read 6,000 statute miles or 5,600 radar miles (7).

To determine the peak transmitter power or receiving sensitivity when one of these factors and the range are given, reverse the procedure. In this case, the antenna gains are subtracted and the system losses are added.

It is convenient to remember that a change of 6 db in the system changes the range by a factor of 2.

HORIZON TRACKERS FOR LUNAR GUIDANCE AND CONTROL SYSTEMS*

K. H. KUHN AND E. W. STARK

Air Armament Division, Sperry Gyroscope Company
Division of Sperry Rand Corporation, Great Neck, New York

That horizon-sensing instruments have a future in space vehicles has already been aptly demonstrated in the TIROS I satellite and rocket tests run as early as the summer of 1959. It is the purpose of this paper to discuss the extension of these techniques to the problem of lunar guidance and control.

In one sense, the lunar problem is somewhat simpler than the problem of the earth horizon-tracker problem due to the lack of an atmosphere near the lunar surface which might obscure the planet-space interface. However, since sightings on the earth will probably also be of interest to a lunar guidance and control system, the problem of the earth horizon tracker is also considered in this paper, including the effects of the earth's atmosphere on the accuracy of the instrument.

A discussion of the oblateness effect, as discussed by Roberson, is briefly reviewed, followed by a detailed discussion of the refraction effects of the earth's atmosphere; it is shown that under certain circumstances, the two effects cancel each other. For high-accuracy instruments, compensations must be introduced. The problem of cloud cover is briefly discussed, and a possible solution, which might be of particular significance to a vehicle re-entering the earth's atmosphere for recovery, is offered. The calculations of radiation, necessary for the design of the instrument, are outlined, including absorption by atmospheric constituents, along with some of the theoretical limitations of the instrument.

The various design considerations, including scan mechanization, choice of detector, signal processing, and other factors, are discussed; however, no attempt is made to offer an optimum design, since this will depend strongly upon the specific mission, the vehicle used, and accuracy requirements. It is hoped, however, that enough information is presented to assist the designer of a guidance and control system in his choice of a horizon-tracking instrument.

*The authors wish to acknowledge the help of Mr. R. Kroeger and Mr. R. Wrobel for their assistance and encouragement during the preparation of this paper, and for their aid in the fulfillment of the contract under which part of the work was accomplished.

The work described was performed under Contract No. AF 33(616)-6674 sponsored by Wright Air Development Division USAF under subcontract from Systems Corporations of America, Los Angeles, California.

THE HORIZON-TRACKING PROBLEM

Horizon-tracking instruments have often been proposed for use in determining the vertical (more accurately, the line of sight from the vehicle to the geometric center of the target planet) on board a space vehicle. A knowledge of the vertical, while necessary, is not sufficient information to establish an attitude reference coordinate system. Another line in space must be established by an independent reference, such as a stellar or inertial one. Other proposals have appeared in the literature which use combinations of stellar, inertial, and gravity references combined with horizon trackers to enhance the performance of the over-all system. It is the purpose of this paper to discuss only the horizon-tracker component, its design features, and the considerations involved in the design of such an instrument.

Other possibilities for the establishment of an attitude reference coordinate system include star sightings, inertial systems, and earth-based radio systems, each with peculiar features, a discussion of which is outside the scope of this paper.

There are at least two different horizon trackers currently available. One is of the circular-scan type; the other, the details of which are not public, is an infrared-sensitive instrument aboard the TIROS satellite. Other proposals have appeared in the open literature by writers representing Barnes Engineering Corporation, Aerojet General Corporation, Sperry Gyroscope Company, Detroit Controls Corporation, I.T. & T. Laboratories, and others.

ANALYTICAL INVESTIGATIONS

Systematic Errors

The accuracy of a horizon scanner depends not only on the particular instrumental errors, which may be of a systematic or random nature, but also on the errors introduced by the geometry of the planet's disc and the physics of its atmosphere. These error-producing factors include:

1. Oblateness of the planet
2. Atmospheric refraction
3. Surface irregularities
4. Cloud cover
5. Atmospheric scattering and absorption

These factors will be treated in turn in the following sections. The oblateness effect and atmospheric refraction are classed as systematic errors, while cloud cover and atmospheric scattering and absorption are treated under radiation calculations. Atmospheric effects will be negligible to a tracker using the moon as a target, due to the lack of atmosphere. However, systems for lunar flight could quite conceivably use horizon-tracker sightings on the earth; for this reason, the effects of the earth's atmosphere will be considered. Surface irregularities are discussed under closed-loop operations.

Range from a Horizon Tracker. A measure of range is available from a horizon-tracking instrument if the diameter of the target planet is known, which is the case for all planets of the solar system. Defining the planet radius as a, the angle subtended at the tracker by the horizon as 2θ, and the distance from the tracker to the planet's geometric center as ρ, it can be shown that

$$\rho = \frac{a}{\sin \theta}$$

By differentiation, the range error can be shown to be related to the error in θ, (the error in measuring the subtended angle) by

$$\frac{\Delta\rho}{\rho} = \Delta\theta \sqrt{\left(\frac{\rho}{a}\right)^2 - 1}$$

and to the error in the planet radius by

$$\frac{\Delta\rho}{\rho} = \frac{\Delta a}{a}$$

so that the percentage in range is equal to the percentage error in knowledge of the target planet's radius.

Figure 1 is a plot of the range error as a function of the ratio for two values of angle-measurement error.

Oblateness Effect. The effect of oblateness has been treated by Roberson [1]. Roberson defines the angle $\Theta(\phi)$ as the subtended angle between the vehicle axis $S_1(a, \beta)$ and the line of sight $S_2(\theta, \phi)$ to the edge of the disc at the true azimuth angle ϕ. $\Theta(\phi)$ is given by

$$\cos \Theta(\phi) = \cos \theta \cos a + \sin \theta \sin a \cos (\phi - \beta)$$

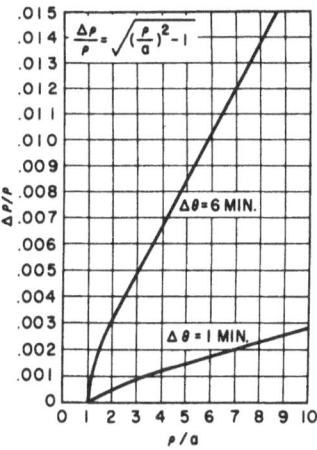

Figure 1. Range accuracy of horizon tracker.

If the vertical has been established almost correctly, a is small and $\Theta(\phi)$ is approximately

$$\Theta(\phi) = \theta - a \cos(\phi - \beta)$$

Roberson further defines the following quantities, formed by averaging

$$\overline{\Theta} = \left(0 \leq \overset{\text{av}}{\phi} \leq 2\pi \right)\Theta(\phi) \qquad \Theta_{34} = \left(\pi \leq \overset{\text{av}}{\phi} \leq 2\pi \right)\Theta(\phi)$$

$$\Theta_{12} = \left(0 \leq \overset{\text{av}}{\phi} \leq \pi \right)\Theta(\phi) \qquad \Theta_{23} = \left(\frac{\pi}{2} \leq \overset{\text{av}}{\phi} \leq \frac{3\pi}{2} \right)\Theta(\phi)$$

$$\Theta_{41} = \left(-\frac{\pi}{2} \leq \overset{\text{av}}{\phi} \leq \frac{\pi}{2} \right)\Theta(\phi)$$

For a spherical planet of radius a, the attitude and position of the vehicle are then

$$\rho = \frac{a}{\sin \theta}$$

$$a \sin \beta = \frac{\pi}{4}\left(\Theta_{34} - \Theta_{12} \right)$$

$$a \cos \beta = \frac{\pi}{4}\left(\Theta_{23} - \Theta_{41} \right)$$

where ρ is the distance from the center of the planet to the vehicle.

Assuming the body has the equation,

$$(1 - e^2)(x^2 + y^2 - a^2) + z^2 = 0$$

where a is the equatorial radius $e^2 = 1 -$ polar radius$^2/a^2$, Roberson finds the following errors due to oblateness:

$$\frac{\partial \rho}{\rho} = \frac{e^2 a^2}{2\rho^2}\left[1 + \left(\frac{\rho^2 - 3a^2}{2a^2}\right)\cos^2 \lambda\right]$$

$$\partial(a \sin \beta) \doteq -\frac{e^2 a^2}{2\rho^2}\sin \lambda \cos \lambda$$

$$\partial(a \cos \beta) = 0$$

where λ is the geometric latitude of the vehicle. The resultant errors are shown in Figures 2 and 3 for the earth.

Refraction Effects. Although the lunar atmosphere will present little if any problem to a horizon tracker using the moon as a target, any appreciable atmosphere will have an effect. This effect could be very significant to a vehicle returning, say, to the earth from the moon. For this reason, the atmospheric refraction is discussed.

Atmospheric refraction will have a very significant effect on the value of Θ as measured from a space vehicle. The optical line of sight S_2 (θ, ϕ), which is tangential to the planetary body, will not be a straight line because of the gradient of index of refraction in the atmosphere. This index gradient yields a continuous curvature of S_2 (θ, ϕ), such that θ, the apparent angle to the horizon, is larger than that expected from simple geometry. θ may be redefined to include the effects of refraction, yielding

$$\theta = \theta_0 + r$$

where r is the atmospheric refraction and θ_0 is the geometric line-of-sight to the horizon. Conversely, the refraction may be considered to produce a change in the apparent size and shape of the planetary disc.

Letting the radius of the oblate planet equal $a + \Delta a(\lambda')$, where λ' is the geometric latitude at which the radius is measured, Δa is given by

$$\Delta a = -\frac{a}{2}\epsilon^2 \sin^2 \lambda'$$

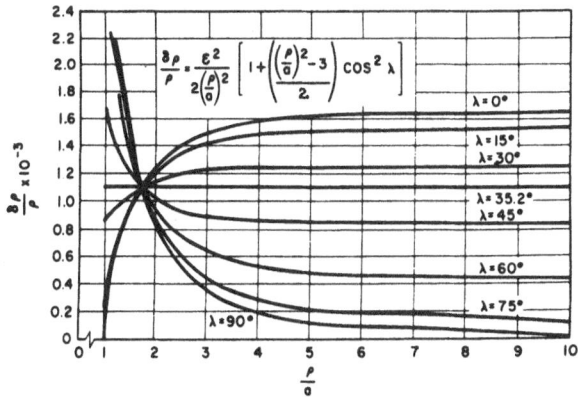

Figure 2. Fractional altitude error (oblateness effect only).

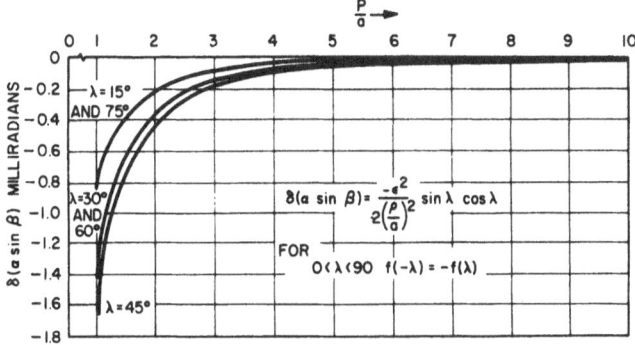

Figure 3. Attitude error.

The atmospheric refraction effect may be found by defining the apparent radius a', which is given by

$$a' = \rho \sin (\theta_0 + r) = \rho(\sin \theta_0 \cos r + \cos \theta_0 \sin r)$$

Since r is small, as is shown below:

$$a' \simeq \rho (\sin \theta_0 + r \cos \theta_0) = a + \Delta a(\lambda) + r(\lambda') \sqrt{\rho^2 - (a + \Delta a)^2}$$

and letting $\Delta a'$ be the effect on a' due to refraction,

$$\Delta a' = r(\lambda') \sqrt{\rho^2 - (a + \Delta a)^2} \approx r(\lambda') \, a \sqrt{\frac{\rho^2}{a^2} - 1}$$

Finally,

$$a' = a \left[1 - \frac{\epsilon^2}{2} \sin^2 \lambda' + r(\lambda') \sqrt{\frac{\rho^2}{a^2} - 1} \right]$$

Defining $r(\lambda')$ as $r_0 + \delta r(\lambda)$,

$$a' = a \left[1 + r_0 \sqrt{\frac{\rho^2}{a^2} - 1} + \delta_r(\lambda) \sqrt{\frac{\rho^2}{a^2} - 1} - \frac{\epsilon^2}{2} (1 - \cos^2 \lambda') \right]$$

λ' in the above equation is the latitude of tangency of ray $S_2(\theta, \phi)$, and neglecting Δa and $\Delta a'$, λ' is given by

$$\sin \lambda' = \frac{\sqrt{\rho^2 - a^2}}{\rho} \left(\cos \lambda \sin \phi + \frac{a}{\sqrt{\rho^2 - a^2}} \sin \lambda \right)$$

$$= \cos \theta \cos \lambda \sin \phi + \sin \theta \sin \lambda$$

Since the term in r_0 is independent of λ' and therefore of λ, θ, and ϕ, its effect may be readily compensated by subtraction of r_0 from θ in the equipment data-processing. An estimate of r_0 must be made and any error in this estimate will affect the accuracy of altitude determination to the same extent as an error in measurement of θ as previously discussed. Error in estimating r_0 will not affect the accuracy of the vertical.

In order to evaluate the effect of $\delta r(\lambda')$ and the magnitude of r_0, an expression for the atmospheric refraction is required.

Chauvenet [2] gives an excellent discussion of the earth's atmospheric refraction based on analysis by Bessel and Argalander, which yields

$$r \approx 0.0109 \left(\frac{P}{P_0} \right)^{1.10} \left(\frac{T_0}{T} \right)^{1.78} \text{ rad}$$

where P_0 is 760 mm or 1013.3 mb and T_0 is 273°K. The refraction for a ray tangent to the surface of the earth, given by the above equation, is dependent on a density distribution, which is exponential with altitude. Since the atmosphere is subject to fluctuations of density distribution, this value of r is subject to variation. However, the accuracy is sufficient to indicate the trend of the refraction effect.

In order to determine $r(\lambda')$, an estimate of the distribution of temperature and pressure with latitude is required. Data from which average temperature and pressure as a function of latitude may be derived are given in [3].

Based on the ARDC atmospheres for arctic winter, arctic summer, model, and tropical conditions, the variation of pressure from standard yields less than 1 per cent change in r. Pressure will therefore be neglected in determining mean refraction.

Three model temperature distributions with latitude are given in Figure 4. These distributions give approximate average temperatures

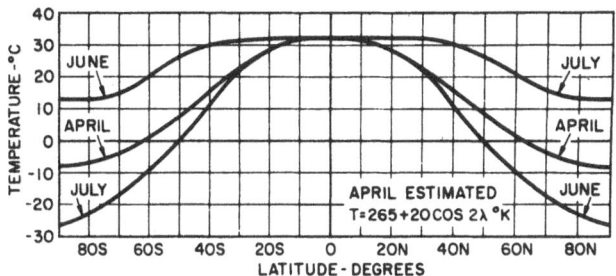

Figure 4. Average temperature distribution.

for July, April, and January. The January and July distributions are based on the data of [3]. The April distribution is an analytical curve, matched to the ARDC models at 40°N and 0° latitude, yielding

$$T_{ap} = 273 + 20 \cos 2\lambda' + 12 = 273 + 40 \cos^2 \lambda' - 8$$

Using the equation for refraction given above, the refraction for the April distribution is

$$r_{ap} = 0.0109 \left[1 - \frac{1.78}{273} (20 \cos 2\lambda' + 12) \right]$$

$$= r_{0_{ap}} + \delta r(\lambda')_{ap} = 10.1 - 1.42 \cos 2\lambda' \text{ mrad}$$

This distribution, as well as the January and July refraction distributions, is shown in Figure 5.

For the April model,

$$\delta r(\lambda') = -1.42 \times 10^{-3} \cos 2\lambda' = -1.42 \times 10^{-3} (1 - 2 \sin^2 \lambda')$$

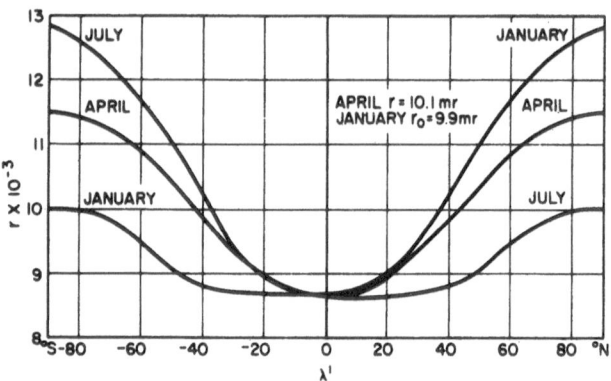

Figure 5. Average refraction distribution with latitude.

Examining the terms of the apparent radius a', which vary with λ', we find

$$\Delta a(\lambda') + \Delta a'(\lambda') =$$

$$a\left[\left(2.84 \times 10^{-3}\sqrt{\frac{\rho^2}{a^2}-1}-\frac{\epsilon^2}{2}\right)\sin^2\lambda' - 1.42 \times 10^{-3}\sqrt{\frac{\rho^2}{a^2}-1}\right]$$

The first term of this function may be defined in terms of a new apparent oblateness factor ϵ, where

$$(\epsilon')^2 = 5.68 \times 10^{-3}\sqrt{\frac{\rho^2}{a^2}-1}-\epsilon^2 = \epsilon^2\left(0.844\sqrt{\frac{\rho^2}{a^2}-1}-1\right)$$

where $\epsilon^2 = 6.73 \times 10^{-3}$. At $\rho/a = 1.55$, $(\epsilon')^2$ goes to zero, and the oblateness and refraction cancel. If $(\epsilon')^2$ is substituted for ϵ^2 in the equations for oblateness error,

$$\frac{\delta\rho'}{\rho} = \left(0.844\sqrt{\frac{\rho^2}{a^2}-1}-1\right)\frac{\delta\rho}{\rho} =$$

$$\left(\frac{5.68 \times 10^{-3}\sqrt{\frac{\rho^2}{a^2}-1}-\epsilon^2}{2\rho^2}\right)a^2\left[1+\left(\frac{\rho^2-3a^2}{2a^2}\right)\cos^2\lambda\right]$$

$$\delta'(a\sin\beta) = \left(0.844\sqrt{\frac{\rho^2}{a^2}-1}-1\right)\delta(a\sin\beta) =$$

$$-\frac{\left(5.68 \times 10^{-3} \sqrt{\frac{\rho^2}{a^2} - 1 - \epsilon^2}\right)}{2\rho^2} a^2 \sin \lambda \cos \lambda$$

$$\delta'(a \cos \beta) = 0$$

These errors for the April model are shown in Figures 6 and 7. The errors shown in the figures are these terms which are a function of λ. In addition, the term of $\delta r(\lambda')$, which is not a function of λ', must be considered. Letting

$$\delta r(\lambda') = \delta r_0 + \delta'r(\lambda') = -1.42 \times 10^{-3} + (2.84 \times 10^{-3}) \sin^2 \lambda'$$

δr_0 may be considered a correction to r_0, such that,

$$r'_{0 \text{April}} = r_{0 \text{April}} + \delta r_{0 \text{April}} = 8.68 \times 10^{-3} \text{ rad}$$

In summary, therefore, the combined effects of oblateness and the refraction resulting from a temperature distribution given by the April model are as follows:

1. The angle θ is increased by approximately 8.7 mrad, independent of latitude. In order to avoid intolerable errors in altitude determination, an estimated correction in θ, as measured by the horizon scanner, must be made in the instrument.

2. The residual fractional altitude error when the above correction is made is given by

$$\frac{\delta'\rho(\lambda)}{\rho} = \left(\frac{5.7 \times 10^{-3}\left(\frac{\rho^2}{a^2} - 1\right)^{1/2} - \epsilon^2}{2\frac{\rho^2}{a^2}}\right)\left[1 + \left(\frac{\rho^2 - 3a^2}{2a^2}\right)\cos 2\lambda\right]$$

In the absence of latitude (λ) data, the best modification which may be made is to correct the measured ρ by the average value of $\delta\rho'$. The residual fractional error is then given by

$$\frac{\delta''\rho(\lambda)}{\rho} = \frac{\delta\rho'(\lambda)}{\rho} - \frac{\delta\rho'\left(\frac{\pi}{4}\right)}{\rho}$$

$$= \left(\frac{5.7 \times 10^{-3}\left(\frac{\rho^2}{a^2} - 1\right)^{1/2} - \epsilon^2}{2\frac{\rho^2}{a^2}}\right)\left(\frac{\rho^2 - 3a^2}{2a^2}\right)\left(\frac{\cos 2\lambda}{2}\right)$$

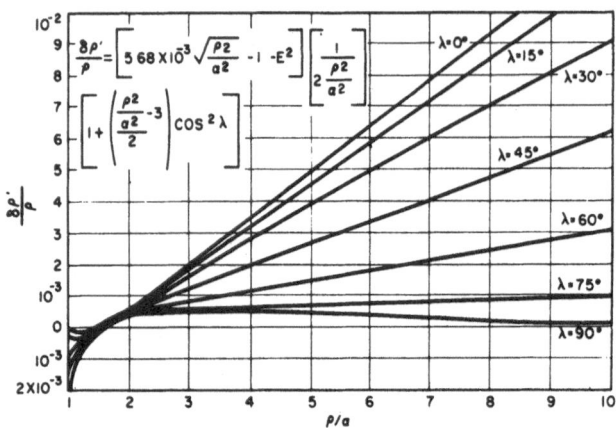

Figure 6. Fractional altitude error with oblateness and refraction.

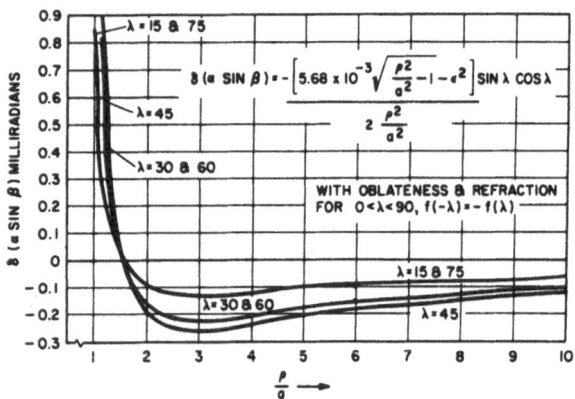

Figure 7. Attitude error.

The maximum fractional altitude error ($\lambda = 0$, 90 degrees) is 1.7×10^{-3} at $\rho/a = 1$, 0 at $\rho/a = 1.55$, and increases thereafter to ∞ as $\rho/a \to \infty$

3. The attitude errors are given by

$$\delta(a \sin \beta) = -\left(\frac{5.7 \times 10^{-3}\left(\dfrac{\rho^2}{a^2} - 1 \right)^{1/2} - \epsilon^2}{2\dfrac{\rho^2}{a^2}} \right) \frac{\sin 2\lambda}{2}$$

$$\delta(a \cos \beta) = 0$$

In the absence of latitude data, no better error can be obtained, as the error changes sign with λ, yielding a zero average.

4. If averaging over an entire orbital period is possible, attitude error will go to zero. The altitude error, however, may be corrected only if the orbit is known.

The error considerations indicated above apply to April through October, when the model-temperature distribution is symmetrical with latitude. For any other time of year, shifts in temperature distribution occur. Based on [3], the extremes which occur in January and July have been shown in Figure 4. For any distribution of interest, an analysis similar to that summarized above will yield the systematic errors. Refraction distributions such as those shown for January and July will have a residual error which is greater than those found for April. The residual error is that error remaining when all constant and $f(\rho/a)$ corrections have been made.

The refraction predicted above is based on visible light. Since the index of refraction of air is a function of wavelength, the refraction must be corrected for wavelength if highest accuracy is required. The index of refraction of air is given by

$$n = 1 + N \times 10^{-6}$$

where

$$N(\underline{\lambda}) = N_{\underline{\lambda}} = \infty \left(1 + \frac{7.52 \times 10^{-3}}{\underline{\lambda}^2} \right)$$

and $\underline{\lambda}$ is given in microns. Examination of the refraction as given by Chauvenet indicates that refraction varies with N in the same manner as with pressure, i.e.:

$$r(\underline{\lambda}) = \left(\frac{N(\lambda)}{N(\underline{\lambda}_0)} \right)^{1.1} r(\lambda_0) \approx \left[1 + 8.27 \times 10^{-3} \left(\frac{1}{\lambda^2} - \frac{1}{\underline{\lambda}_0{}^2} \right) \right] r(\underline{\lambda}_0)$$

The value of $\underline{\lambda}_0$ used in the analysis above was 0.587 micron. The maximum value of this correction, as wavelength increases from 0.587 micron, is 2.4 per cent.

Variation of refraction from the predicted value, which may reach 0.2 mrad, will introduce noise in θ (ϕ) as measured by the scanner. A treatment of this effect is beyond the scope of the present work. Additional analysis, based on statistical temperature fluctuations from the mean model assumed here, would yield data on this noise.

The fluctuation of refraction from the predicted value may be the limiting factor on the accuracy obtainable with horizon-scanning equipment. The fluctuation may be considered to have two components — (1) the average deviation of $r(\phi)$ over $0 \leq \phi \leq 2\pi$ from the mean assumed, and (2) the components which vary with ϕ. The latter terms, which affect attitude but not altitude, will be reduced by the averaging process discussed above. The analysis would follow, in general, the outline given in this report for analysis of the effects of surface irregularities.

A similar analysis of refraction for the atmospheres of the other planets is necessary to determine the correction factors required. Temperature, pressure, and/or density models, together with index or refraction data, for any atmosphere will allow computation of the refraction effect.

Radiation Calculations

Source Energy. All planets radiate energy in the infrared region of the electromagnetic spectrum in addition to reflecting sunlight in the visible region of the spectrum. It is the purpose of the following discussion to survey these characteristics for properties suitable for horizon-scanning. The particular bodies of concern are the earth and moon; the regions of the electromagnetic spectrum of interest are as follows:

1. The ultraviolet and visible (0.2 to 0.7 micron)
2. The infrared (0.7 to 30.0 microns)
3. The far infrared (near 1,000 microns)

Three basic properties affect the radiation characteristics available for use by a radiation-sensing instrument. These are

1. The planet's temperature and emissivity
2. The planet's reflection of solar radiation
3. The characteristics of the planet's atmosphere

Table 1 lists the characteristics of interest extracted from [4] and [5].

The thermal radiation from the surface of a planet may be described in terms of the product of the radiation from a blackbody at the temperature of the surface, and the effective graybody emissivity

TABLE 1. THERMAL AND GEOMETRIC PROPERTIES OF PLANETS

Mean Diameter	Earth	Moon	Sun
KM	12,742.46	3476	1.393×10^6
NM	6,875.74	1877	751,680
Equatorial diameter (2a)			
KM	12,756.78	−	−
NM	6,883.69	−	−
Polar diameter $2a\,(1-f_p)$			
KM	12,713.82	−	−
NM	6,860.50	−	−
Flattening polar f_p	1/297	1/2000	−
equator f_p	0	1/1667	−
Mean distance from sun			−
KM	149.5×10^6	384,400*	−
NM	80.7×10^6	207,400*	−
AU	1.00000	−	−
Maximum distance	−	406,700* km	−
AU	1.016727	219,500* nm	−
Minimum distance	−	356,400* km	−
AU	0.983273	192,500* nm	−
Visual albedo	0.39	0.072	−
Surface temperature	−	−	−
Maximum °K	330	373	−
Average °K	290	274	−
Minimum °K	205	120 (?)	−
Radiation temperature	−	−	−
Block sphere °K	279	−	5750° total
Maximum °K	349	−	7000°K (i-r)
Average °K	246	−	6100°K (visible)

*From earth

of the surface. The total radiant emittance of a blackbody is given by the Stefan-Boltzmann equation,

$$W_{bb} = \sigma T^4$$

where

W_{bb} = total radiant emittance (w-m^{-2})

σ = Stefan-Boltzmann constant = 5.6686×10^{-8} (w-m^{-2} – °K^{-4})

T = temperature (°K)

The effective graybody emissivity, ϵ_{gb} is the ratio of the radiant emittance of the surface, W_s, to W_{bb} defined above. For many cases, emissivity is a function of wavelength and temperature, however, over the temperature range of interest, effective emissivity ϵ_{gb} is nearly constant.

When the spectral distribution of radiant emittance is significant, the blackbody spectral radiant emittance must be computed from Planck's distribution law:

$$W_{bb}(\lambda,T) = C_1 \lambda^{-5} \left(\frac{C_2}{e^{\lambda T} - 1} \right)^{-1}$$

where

$W_{bb}(\lambda,T)$ = spectral radiant emittance in the spectral region
λ to $\lambda + d\lambda$ (w-m^{-2}-micron^{-1})
$C_1 = 3.7413 \times 10^8$ (w-m^{-2}-microns^4)
$C_2 = 1.4388 \times 10^4$ (micron-$^\circ$K)
λ = wavelength (microns)
T = temperature ($^\circ$K)

The surface spectral radiant emittance is then

$$W_s(\lambda,T) = \epsilon_s(\lambda) W_{bb}(\lambda,T)$$

where

$\epsilon_s(\lambda)$ = emissivity of the surface at wavelength λ

By differentiating the Planck equation with respect to wavelength, the wavelength of maximum radiant emittance can be derived. The resulting equation is the Wien displacement law:

$$\lambda_m T = 2891 \text{ micron} - K^\circ$$

where

λ_m = wavelength of maximum radiant emittance in microns
T = temperature in degrees Kelvin

A plot of the Wien displacement law in the temperature range of interest is shown in Figure 8. Note that the figures show that the wavelength of peak emittance for all the bodies of interest falls between 8 and 24 microns, or if the cold side of the moon is ignored, all wavelengths of peak emittance fall between 8 and 15 microns.

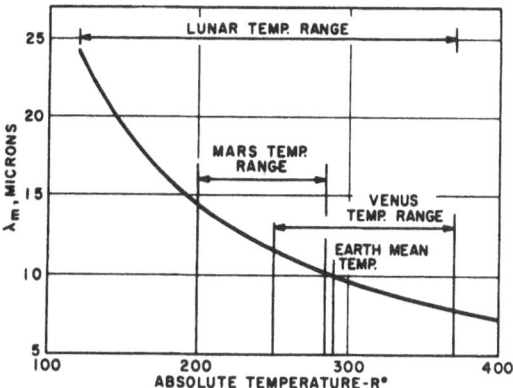

Figure 8. Wavelength of peak emittance

The solar irradiance incident on a planet is nearly blackbody in character, with an effective temperature of approximately 5,800°K. The spectral irradiance $H(\lambda)$ may be computed from Planck's equation modified by a constant to account for the geometric factors. The wavelength of peak irradiance can be computed, on the basis of solar temperature, from the Wien displacement law. This computation yields a wavelength of maximum irradiance of 0.5 micron.

The radiant emittance resulting from the reflection of solar radiation is given by the product of the solar irradiance and the reflectance ρ of the surface. This ratio of reflected to incident radiation is also known as the albedo, although the definition of *albedo* is quite loose. As was true for emissivity, radiant reflectance may vary as a function of wavelength. The spectral radiant emittance is then given by

$$W(\lambda) = \rho(\lambda) H(\lambda)$$

For the problem considered here, $H(\lambda)$ varies more rapidly with wavelength than $\rho(\lambda)$, and the reflected radiation will peak in the visible region of the spectrum.

Thus two distinct bands of interest are defined, one for thermal radiation, which peaks in the region from 7 to 25 microns in the infrared, and the other for reflected solar radiation, which peaks in the blue region of the visible spectrum.

Atmospheric Phenomena [6,7].   The discussion thus far has neglected the important effects of the planetary atmosphere present in every case of interest with the exception of the moon. The atmosphere will have a number of absorption bands, which attenuate radiation being transmitted through the atmosphere. Since emissivity equals absorptivity, the atmosphere will reradiate energy as a selective radiator, with a spectral radiant emittance equal to that of a blackbody at atmospheric temperature, multiplied by the spectral absorptivity of the gas. The radiation from a segment of the path will then be attenuated and reradiated by the atmosphere for the remainder of the path. This absorption and reradiation will produce marked changes in the spectral emittance for both thermal and reflected solar radiation.

In addition, planetary atmospheres produce scattering of radiation, most evident in clouds. The scattering function is less spectrally selective and may totally prevent transmission of radiation in the visible and infrared regions of greatest interest. The most strongly scattering constituents also will radiate nearly as a blackbody.

If either absorption or scattering is sufficiently intense, the radiation from the surface or lower atmospheric regions will be negligible compared to that received from the higher altitudes. In that case the perceived horizon will be an atmospheric one. If the attenuation is less severe or occurs over a lengthy path, the effect will produce a "fuzzy" edge to the apparent planetary disc. This reduced edge gradient may affect the accuracy of a horizon sensor.

The basic attenuation phenomena of the atmosphere are discussed below. Detailed analysis of the atmospheric effects on horizon-scanning may be found later in this section.

If, for the particular application, an atmospheric horizon sufficiently approximates the true horizon, the sensing system can be designed to take advantage of an absorption band of an atmospheric constituent gas, thus capitalizing on what, under other circumstances, could be a severe disadvantage

The factors affecting atmospheric attenuation and radiation, may be summarized as follows:

1. Scattering by haze, smog, or dust particles
2. Scattering by the water droplets, constituting fog, clouds, and rain
3. Absorption by the water droplets, constituting fog, rain, and clouds
4. Absorption by the atmospheric gases

These attenuations are superposed on the normal inverse-square effect. The general effect of scattering is to remove the desired radiation from the line-of-sight path and to introduce extraneous or unwanted radiation into the line of sight. Absorption is an energy transferral process; absorption and emission are complimentary processes in every gaseous medium.

Scattering [7]. The two important factors determining the attenuation caused by scattering are the ratio of the mean particle radius to the wavelength of the radiation, and the number of particles along the sighting path. The transmission decreases exponentially according to:

$$T = e^{-bR}$$

where

R = length of path through the scattering medium

$b = \pi a^2 N K_s$

with

a = mean particle radius

N = number of particles per unit volume

K_s = scattering cross section of the particle

The K_s term is a measure of the scattering efficiency of the particle and depends in a complex manner on a factor α, which in turn is given by

$$\alpha = \frac{2\pi a}{\lambda}$$

where

α = scattering coefficient

a = mean particle radius

λ = wavelength of the radiation

The relationship between K_s and α is illustrated in Figure 9. When the particle radius is large compared to the wavelength, α is large; K_s then has a value of about two, and the scattering shows little wavelength dependence. When the particle radius equals the wavelength, K_s attains a maximum value and the particle has the greatest effectiveness as a scatterer. When the radius is small, K_s decreases rapidly, and in this region varies inversely as the fourth power of the wavelength. This is the region of Rayleigh scattering postulated by Lord Rayleigh to explain the blue color of the sky in terms of molecular scattering or sunlight.

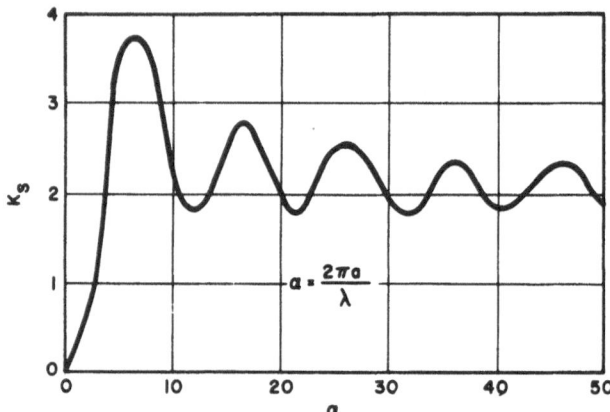

Figure 9. Scattering coefficient.

The sizes of haze particles vary widely, being generally dependent on relative humidity, but a radius of 0.5 micron may be considered
as a typical mean value. Thus a considerable improvement in transmission can be obtained by even a slight increase of the operating
wavelengths beyond visible light. For atmospheric haze, smog, and
dust, the particles are numerous, but sufficiently small that the
infrared wavelengths from 0.8 to 15.0 microns have good transmission
capabilities. With respect to fog and clouds, the situation is much
worse because the particles are larger. Typical size distributions
are shown in Figure 10.

Assuming a uniform cloud composed of droplets of 10-micron
mean radius and a concentration of 200 droplets/cc, the computed
transmission for radiation of various wavelengths as a function of
path length is plotted in Figure 11. Although this is a rather modest
cloud, it can be seen that the transmission of the shorter i-r wavelengths are effectively reduced to zero for path lengths greater than
100 m.

Far Infrared Radiation and Absorption by Water. If it becomes
necessary to sense the true horizon at all times, such as during a
re-entry maneuver, the effects of scattering by atmospheric particles
become all important. When the wavelength is great compared to the
particle size, the scattering varies inversely as the fourth power of
the wavelength. Haze particles, having sizes of 1 micron or less
and usually found only at the lower altitudes, probably will not have
a serious effect on the longer wavelengths considered here. Infrared,

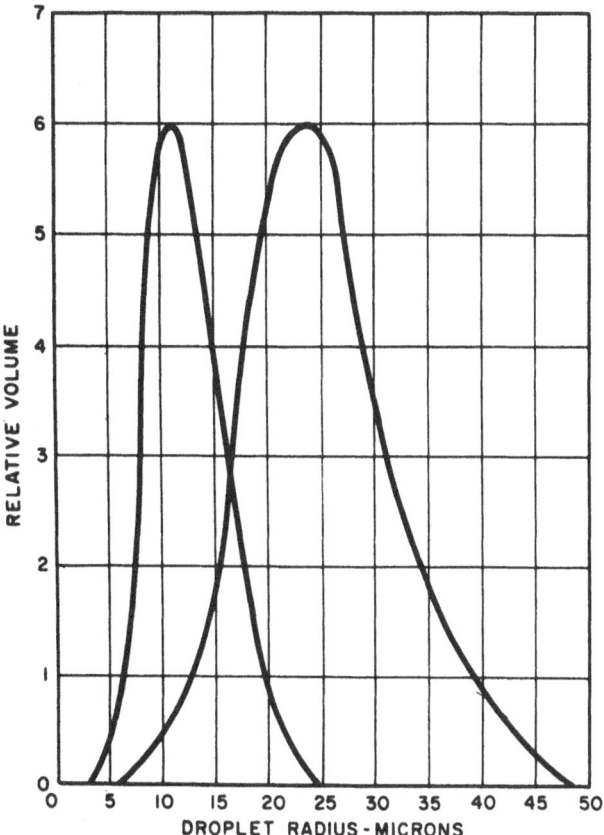

Figure 10. Droplet size distribution for cloud and fog.

even at the shorter wavelengths, is known to have a pronounced haze-cutting ability. Fog and cloud particles, however, have diameters ranging from 20 to 60 microns or so. Considering the great concentrations of such large scattering particles as will be encountered along the slant paths through fog banks and cloud cover, it appears that little reliance can be obtained from these wavelengths for penetrating fog and clouds. The situation is aggravated by raindrops and snowflakes which have sizes of 1 mm and greater, although their concentrations are less.

Some hope can be held for reasonable all-weather capabilities in going to the very long infrared wavelengths of 1 mm (1,000 microns)

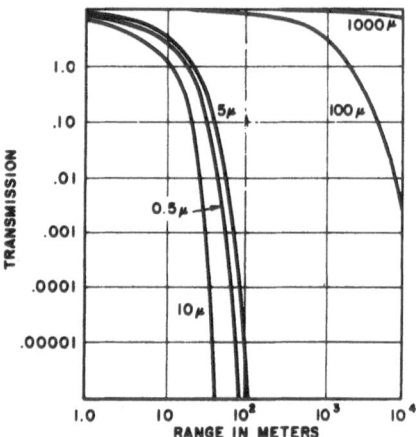

Figure 11. Transmission of various wavelengths through typical cloud.

and greater. Several good transmission windows do exist, and the wavelengths exceed the sizes of fog and cloud particles by about two orders of magnitude. Although there is much less energy being radiated at these wavelengths, the earth-sky contrast is thereby not necessarily made worse. Even the signal-to-external-noise ratio may be improved. Little work has been done on the detection of these long waves which border on microwaves, and to date no practical applications seem to exist. We wish to call attention to this unused portion of the spectrum and to suggest its application to horizon-scanning. A discussion of this application is given in [8].

There remains the problem of absorption by the liquid-water content. The absorption is caused by the fact that water is a lossy dielectric at infrared and millimeter wavelengths. The absorption is given by:

$$r = (0.44/2)M \text{ db/km}$$

where

    r = attenuation by liquid-water absorption
    M = water content of the fog or cloud in grams per cubic meter

Except in the heaviest sea fogs, the maximum value of M which is likely to occur is about one, and usually it is much less. The maximum possible path-length due to the height of the atmosphere when scanning the horizon from a satellite is of the order of 370 km; the

maximum likely path-length containing liquid water of the order of
100 km, due to the limited altitude of water-bearing clouds. Under
these assumptions the maximum attenuation to be expected is of the
order of 20 db, which is appreciable but not sufficient to rule out
millimetric techniques for all applications.

Absorption by Atmospheric Gases. As is discussed below, absorption and emission by atmospheric gases is a function of composition, concentration, pressure, temperature, and density of the
absorbent and the atmosphere. The composition of the atmospheres
are presented below for the planets under consideration.

Earth Atmosphere (permanent, constituent) [5]

Constituent	% by Volume	% by Weight	Reduced Thickness atm-cm NTP
Nitrogen (N_2)	78.088	75.527	624,343
Oxygen (O_2)	20.949	23.143	167,495
Argon (A)	0.93	1.282	7,435.69
Carbon dioxide (CO_2)	0.03	0.0456	239.861
Neon (Ne)	1.8×10^{-3}	1.25×10^{-3}	14.3817
Helium (He)	5.24×10^{-4}	7.24×10^{-5}	4.18957
Methane (CH_4)	1.4×10^{-4}	7.75×10^{-5}	1.11935
Krypton (Kr)	1.14×10^{-4}	3.30×10^{-4}	0.911472
Nitrous oxide (N_2O)	5×10^{-5}	7.6×10^{-5}	0.399768
Xenon (Xe)	8.6×10^{-6}	3.90×10^{-5}	0.068760
Hydrogen (H_2)	5×10^{-5}	3.48×10^{-6}	0.399768

The above listing for earth can be considered invariant up to an
altitude of 90 km.

The important variable gases in the earth's atmosphere are
ozone (O_3) and water vapor (H_2O). Ozone appears only in a band in
the upper atmosphere, and is subject to variation with latitude and
with season. Water vapor is found principally in the lower 10 km of
the atmosphere, where it may at times constitute 2 to 3 per cent by
weight of a given volume of the atmosphere at that level. The actual
amount of water vapor in any portion of the atmosphere depends on
physical location, temperature, and weather conditions; it is therefore extremely variable.

Absorption Bands. Absorption by the atmospheric constituent
gases and vapors occurs at discrete wavelengths which are associated with the electric dipole moment of a particular molecule.

In the atmosphere of the earth, there are three major absorbents. These are water vapor, carbon dioxide, and ozone. Water vapor is variable in concentration and generally has its effect at lower altitudes. Its major absorption bands in the infrared spectrum are centered about 6.3, 2.7, 1.87, and 1.38 microns. Carbon dioxide differs from water vapor inasmuch as its concentration is reported to be constant up to at least 90 km. The major absorption bands of CO_2 are centered at 15.0, 4.3, 2.7, and 2.0 microns. Ozone is not normally present in the atmosphere outside of the ozonosphere, i.e., that band of ozone that surrounds the earth at altitudes from 10 to 40 km, with both the altitude and concentration subject to diurnal and seasonal variations. Ozone presents a strong absorption band at 9.6 microns in the infrared.

Of particular interest is the ultraviolet absorption band of ozone between 0.2 and 0.3 micron (maximum absorption at 0.26 micron). At its maximum density of about 50,000 to 100,000 feet, the ozone layer has a thickness of about 0.3 cm when reduced to *NTP* conditions; this layer forms a powerful heat reservoir which maintains approximately constant temperature in the stratosphere. Diurnal variations in O_3 content indicate higher content at night; seasonal variations show a maximum in spring and minimum in autumn, with the greatest seasonal variation at the higher latitudes [5]. Terrestrial magnetic disturbances affect the ozone content, but apparently the sunspot cycles do not.

Figure 12 presents the total solar spectrum as seen through

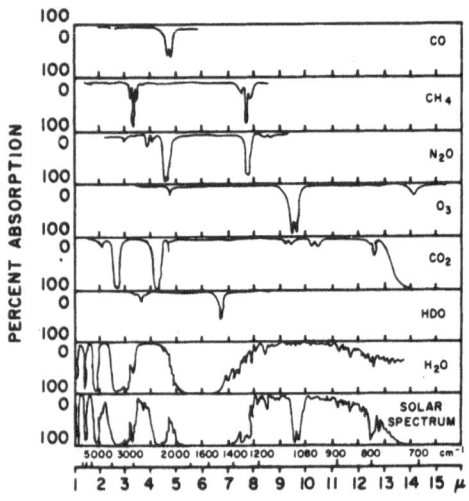

Figure 12. Solar spectrum.

earth's atmosphere and also shows the absorption bands of the three major and several minor absorbing constituents.

Absorption Band Models. The absorption of energy by the atmosphere can be theoretically approximated by the use of various absorption models. Absorption occurs in finite bands which are a grouping of related lines. The lines in a band are broadened because of three general causes: (1) natural broadening, independent of all external effects; (2) Doppler broadening, due to the thermal motion of the molecules; and (3) broadening due to the influence on the absorbing or emitting molecules by other molecules (pressure broadening) or by electric or magnetic fields.

The Elsasser model for an idealized absorption band consists of an infinite array of equally intense, equidistant lines, each of the Lorentz₁ shape with the same half-width. The resulting average absorption of a band can then be represented by

$$A = \text{erf}\left(\frac{\ell w}{2}\right)^{1/2}$$

where

    A = absorption
    ℓ = generalized absorption coefficient as defined in the reference
    w = absorbent concentration

Conversely, Goody has considered in his model the case of a completely disordered band, i.e., the so-called random or statistical model. It can be shown for the Goody model that the average fractional transmission is:

$$\overline{T}[a,\omega,\sigma] = \exp\left(\frac{-\omega\sigma a}{\delta\left(a^2 + \frac{\omega\sigma a}{\pi}\right)^{1/2}}\right)$$

where

    $\overline{T}$ = average fractional transmission in the interval considered
    ω = absorbent concentration
    a = half-width of the lines in the band
    σ = mean line strength
    δ = mean line spacing

The model that should be used depends on the particular band structure. For instance, the Elsasser model fits the 15-micron CO_2 band, while the Goody model is applicable to the H_2O bands.

In many cases, even the Elsasser and Goody models contain more detail than is desirable. When broad spectral regions are considered, such that the entire absorption band is within the region of interest, the desired absorption parameter may be $\overline{A}$, the total fractional absorption of the band, where

$$\overline{A} = \frac{1}{v_2 - v_1} \int A_v dv$$

and

$\overline{A}$ = absorption over a band
v_1, v_2 = frequency limits of the band
A_v = incremental absorption

Howard, et al., have determined empirical equations for this absorption for the CO_2 and H_2O bands by means of artificial atmospheric paths. The first equation is

$$\int A_v dv = c\omega^{1/2} [P + p]^K$$

and is applicable to a weak band, which is considered to consist of many lines far enough apart that the effects of overlapping can be ignored.

The strong-band fit is described by

$$\int A_v dv = C + D \log \omega + K \log [P + p]$$

where

$c, k, C, D,$ and K are constants dependent on the band
ω = concentration of the absorbent
P = total pressure of the medium
p = partial pressure of the absorbent

The applicable constants are presented in the reference. The constants are a function of the band limits chosen. Howard, et al., have chosen empirical limits such that the entire absorption band is within the band limits.

Application of Model Absorption Band Data. For the horizon-scanner problem, the final information required is that the distribution of radiance seen by the scanner as θ — the angle between the line of sight and the vertical — varies. In particular, the horizon gradient of radiance is required. In the region $0 \leq \theta \leq \sin^{-1} a/\rho$, the radiance may be found as follows:

1. Compute the radiance of the planetary surface by means of Planck's radiation law based on surface temperature and emissivity.

2. Compute the atmospheric transmission as a function of θ for the slant path from the surface to the vehicle. Howard, *et al.*, give references by King, Elsasser, Goodson, and Plass on methods of applying constant-pressure absorption data to oblique atmospheric paths. Birch (*IRIS*, Vol. 2, No. 2) gives a solution for the 4.3-micron CO_2 band suitable for digital computer calculations.

3. Compute the radiance of the atmosphere as seen from the vehicle. Since the emissivity of the atmosphere equals its absorptivity, the radiance from a layer at a given temperature may be computed by means of Planck's law and absorption data. This radiance must then be modified by the absorption from the layer in question to the vehicle and the results integrated over the entire path.

4. The net radiance seen from the vehicle is then found by multiplying the radiance from step (1) by the transmission from step (2) and then adding the radiance found from step (3).

In the region $\sin^{-1} a/\rho < \theta < \pi$, the line of sight does not end at the planetary surface. The radiance is therefore that due to the space background modified by the transmission through the atmosphere along the line of sight plus the atmospheric radiance. This is found in the same way as for $\theta < \sin^{-1} a/\rho$, with the exception that the pressure and temperature of the atmosphere along the path ranges from the free-space minimum through a maximum at the point of closest approach to the planetary surface to the free-space minimum beyond the atmosphere.

Computation of these radiance functions must be done graphically or by numerical integration. For the earth, the literature provides sufficient data on the parameters involved. In the case of the atmospheres of the other planets, the atmospheric absorption constants have not been determined in a form readily adaptable to the required computations.

Lunar Atmosphere. For all practical purposes, and particularly for horizon-scanning, the moon is considered to have no atmosphere. The pressure is less than 10^{-5} atm.

Theoretical Limitations

Theoretical limits on the performance of horizon-tracking instruments are imposed by three fundamental phenomena:

1. Target temperature, imposing primarily a range limitation
2. Optical design, limiting angular accuracy
3. The radiation characteristics of the horizon

Target temperature defines the level of the target radiant energy, and hence, signal-to-noise ratio at a given range. Since some minimum signal-to-noise ratio must be defined, and some maximum allowable aperture is defined by the maximum allowable physical dimensions of the instrument, this is essentially a range limitation. However, this limitation is not effective until the tracker is several planetary diameters away from the target planet horizon. Range limitations are discussed in [9] and [10].

Angular accuracy, on the other hand, is limited by the performance of the optical system. Chromatic aberrations, in the case of refractive optics, can limit the image resolution as can the diffraction in the image plane if the optical system is not carefully designed. These errors are discussed in detail in [11].

The radiation gradient is defined as the rate of change of radiation intensity with respect to angle swept through as the field of view is traversed across the edge of the target planet's apparent disc. If this gradient is defined by the true planet-space interface, the gradient is very sharp and the only sources of errors are the terrain features (mountain peaks, valleys, etc.), which could become large contributors at low altitudes, and the oblateness effect previously described.

If the gradient is not defined by the planet-space interface, but is also a function of the characteristics of the atmosphere, the error sources are somewhat different. In this case, two possible error sources are present; if the instrument is dependent on the planet-surface interface and must look through the atmosphere, atmospheric refractive effects, which may be a function of position on the surface (as previously described for the case of the earth's atmosphere), or local weather conditions, may become a fundamental limitation. The other possibility is concerned with the case where the horizon is not defined by the planet-space interface, but by some characteristic of the atmosphere such as an absorption band of an atmospheric constituent, in which case the limitation is defined by the correspondence between the atmospheric horizon and the geographic or geodetic horizon, whichever is of interest.

No measurements on the atmospheric-horizon gradient are available in the open literature, but estimates can be made if the temperature of the atmosphere, partial pressure of the particular atmospheric constituent and the absorption characteristics of the constituent are known.

In rudimentary, nonimage-forming systems, the angular accuracy may be defined by the instrument field of view, particularly if little is known of the characteristics of the intensity gradient as the field crosses the horizon. The field of view is defined by the angle subtended by the detector at a distance equal to the focal length away from the focal point of the optical system. In a rudimentary system, the only available information would be the change in detector output as this narrow field traverses the horizon. The angle between the optical axis and a reference can be measured as the nominal angle of interest, and the field of view then is the error region. Figure 13 illustrates this concept.

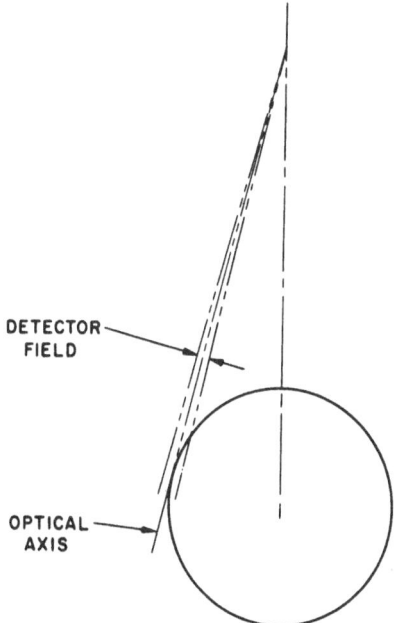

Figure 13. Angular error.

DESIGN CONSIDERATIONS

Scanning of the Horizon Disk

Fundamentally, scanning techniques may be grouped in four broad classes:

1. Static systems, in which the target is examined against a star or space background, and an image formed and possibly scanned electronically [12].
2. Circular-sweep systems, in which the optical system is rotated through a complete circle and information derived from the portion of the circle that intercepts the horizon [13].
3. Discontinuity trackers, in which the optical system tracks the target edge discontinuity [14].
4. Sector-sweep systems, in which the optical system is oscillated across the field of view presented by the target disk, but not outside this field, except possibly for target search and acquisition.

This grouping is not all inclusive, but most types of scanning with few exceptions can be included in one of the four major classes. The basic features of each type of system are indicated in Table 2.

Presumably, any of the four classes of scan techniques can be used with any type of detector. The only limitations of scanning speeds and, hence, information rate are those imposed by the detector or momentum considerations; there are no speed limitations imposed by the scanning technique. Static systems which operate in the near-infrared region (near 3 microns) have been described in the literature; apparently there is no obvious reason why a system could not be designed for a region near 10 or 20 microns (which would be the region of interest for a satellite or space vehicle application) if a system of this class can be shown to have advantageous features. The principal problem is one of finding a semiconducting material that will pass energy at 10 or 20 microns wavelength and change its absorptivity at this wavelength when struck by an electron.

Discontinuity trackers — descriptions of which have appeared in the literature — depend upon (1) a field of view moving across the horizon discontinuity in a small amplitude, and (2) relatively high frequency oscillation superimposed on a conical motion of the field. The apex angle of the cone is dependent on the angle subtended by the horizon disk. This type of tracking involves a fairly complicated mechanical system to generate the required field motion, but this

TABLE 2. COMPARISON OF SCANNING TECHNIQUES

	Advantages	Disadvantages
Static systems	No moving parts; fast image scanning, limited only by persistance of screen	Long-image tube required; no range data available; current work includes only short i-r wavelengths; development required to get best results in longer wavelengths; extra complication to get range data; high scan voltage required
Circular-sweep systems	Mechanically and electronically simple; low-power-drain current available	Two separate tracking heads required; considerable time wasted while optics scan space instead of horizon; optical system must protrude from vehicle
Discontinuity trackers	Continuous data available; target always in field of view; range data readily available	Complicated scan pattern must be generated in some systems, but this might be eliminated by proper reticle design
Sector-sweep systems	Target always in view; range data readily available	Continuous oscillating transfer of momentum to vehicle; mechanically complicated

problem conceivably could be eliminated by proper modulation reticle design.

Since it is theoretically possible to discriminate between a given target and a given background by designing a reticle pattern which moves relative to the projected image of the target and background, it should also be possible to discriminate between the horizon discontinuity and the constant signals representing the target planet and free space. The result will be a time-modulated signal representing the horizon, with little or no signal from the planet or free space.

A simple example of this type of discrimination follows: if a picket fence were scanned by a line slit oriented parallel to the pickets, but moving in a direction perpendicular to the length of the pickets, a large modulated signal would result due to the change in intensity as the slit alternately covered and uncovered the pickets. However, if the slit were perpendicular to the pickets, no signal modulation would result, because the same amount of pickets and spaces would be included in the line slit at all times during the scan. It appears that a modification of this scheme could be made to detect a discontinuity, such as that presented by a horizon. Detailed analyses of this type of filtering are provided in [15] and [16].

Another broad class of systems (not included in the original four classes) involves the use of multielement detectors, which can relieve or eliminate the scanning problem [17]. In this type of system, a group of detectors views the horizon in some sort of restricted scan motion or from a fixed, nonrotating orientation. This class of instruments suffers from the suspect reliability of more than one detecting element and also from the fact that, in almost all cases, this type of instrument depends on a measurement of the radiant intensity rather than the intensity gradient. Hence, the spectral response of the detectors must be accurately matched, and this may limit the accuracy of the instrument.

Another broad class of instruments involves a fixed detector with refractive optical elements used to scan the incoming radiation, thus sweeping the field optically. Refractive systems of this nature are useful in the visible region, but are of limited value in the infrared spectrum, because the signal radiant power levels are usually much lower than in the visible region, and hence the optical losses due to absorption in the optical element may be prohibitive.

Thus, six classes on instruments are defined, each with peculiar advantages and disadvantages. The choice of the classes of system can be based to some extent on the information requirements, such as vertical indication, range, continuous or intermittent data, and information rate. Beyond these features, the choice is one of instrumentation convenience to the particular mission or vehicle under consideration.

The effects of surface irregularity, oblateness, atmospheric refraction, and cloud cover may all be reduced by averaging the value of $H(\phi)$, as defined by Roberson, over $0 \leq \phi \leq 2\pi$ and $(n-1)\,\pi/2 \leq \phi \leq (n+1)\,\pi/2$ where $n = 0, 1, 2, 3$. For the greatest accuracy, this averaging is required. In addition, as discussed in connection with atmospheric refraction, certain errors may be predicted and compensated for. Both averaging and compensation place certain restrictions on scanner design and signal-processing.

The discontinuity-tracker class of horizon scanner lends itself to the averaging requirement, since the scan variable is ϕ. This class may therefore be the least complex, providing the requirements for discontinuity tracking, as opposed to other techniques, are not unduly complex in themselves.

The circular and sector-sweep systems are similar with the exception that the sweep angle is limited to less than π rad in the sector scan, while the circular scan covers 2π rad. In order to allow averaging, the plane of the scan must be rotated in azimuth angle ϕ. This rotation eliminates the requirement for two scanners in perpendicular planes, but adds an additional requirement for an azimuth motion. If this motion is available as a spin of the vehicle about the vertical, the circular or sector sweep class of horizon scanner becomes very attractive. If the vehicle does not spin about the vertical axis, this motion must be provided by the horizon scanner itself. The disturbance torques created by this additional scan motion must be considered.

The static class of horizon scanner yields an image which contains all values of ϕ continuously. The ease of averaging will depend on the ingenuity with which the data contained in the image is processed. While the static horizon scanner is mechanically the simplest, the signal-processing required is more sophisticated and may result in undesirable complexity.

Detector Choice

Spectral Requirements and Sensitivity. The surface-temperature range of the two celestial bodies of interest to this paper (earth and moon) lie in the range of 120° to 380 °K. Their significant blackbody thermal emissions, therefore, cover the range from 3.5 microns to beyond 100 microns, with their peak emissions varying from 25 to 7.8 microns respectively. Ninety per cent of the energy over the 200° to 380 °K range, which applies to the earth but not the moon, is emitted between 6 to 50 microns. The moon has the only surface temperature in this study as low as 120 °K, obtained at its unilluminated far side. However, even at this lower temperature limit, 60 per cent of the total emission falls within 40 microns. Hence, a detector covering the 5 to 40 micron range should prove perfectly adequate, and even a long wavelength limit of about 30 microns would doubtless be satisfactory if the detector with this limit were

superior in other respects. Any narrow wavelength band of sensitivity chosen to maximize system accuracy in the presence of an
atmosphere must necessarily lie within the region of significant
blackbody radiation.

Detectors suitable for this range include the thermal detectors,
such as Golay and other pneumatic types, thermistor bolometers,
and high-sensitivity thermopiles. Semiconductor (extrinsic) photoconductive cells are now being used in this range as they become
available and practical; however, only a few adequately cover the
spectral range under discussion. Also this class of detectors ex-
hibits sensitivities which vary significantly over their useful ranges,
as seen in Figure 14. On the other hand, all of the thermal types
mentioned have essentially flat responses from the visual spectrum
to the millimetric region, and since another source of important and
useful emission from the planets is their reflected solar radiation,
the visual spectra also should be included in a flexible attitude
sensor system.

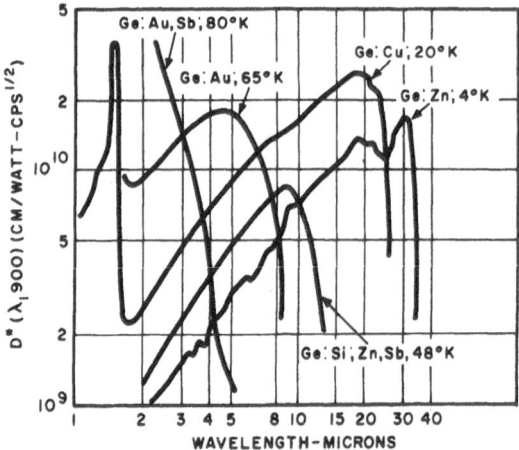

Figure 14.   Extrinsic photoconductive detectors.

If then an i-r photoconductor were used, a separate photoemissive
detector might be needed to include the 0.3 to 1.0 micron range. If
sufficient solar energy exists in the 1.0 to 2.0 micron region — as it
would in most cases — the intrinsic response of the photoconductor,
which occurs at about 1.8 microns for germanium types, can be used
to detect in this portion of the spectrum. Note that in Figure 14 that

this part of the Ge : Cu detector response peaks up at between 1.6 and 1.8 microns and has higher detectivity than at its peak extrinsic response at 20 microns. These peaks exist for the other types but were not shown on this plot. The intrinsic peak for the Perkin-Elmer Ge : Zn detector occurs at essentially the same wavelength and about the same maximum D^* as for Ge : Cu. On the other hand, a Golay detector completely covers the range of 0.5 to 40 microns. (The longwave limit is merely a question of i-r window material and can be further extended, using another, or combination, of window materials.) The sensitivity of the Golay detector is of the same order as the best state-of-the-art photoconductors.

Table 3 shows a comparison between the most sensitive detectors in the 5- to 40-micron range, indicating sensitivities in noise equivalent power and D^* averaged only within the important 8- to 14-micron region (except for the thermal detectors). As seen from Figure 14, which shows detectivity D^* versus wavelength for the most representative photoconductors in this range, only the copper- and zinc-doped germanium types cover this spectrum at all adequately. Table 3 also gives the time constant, effective aperture area, and cooling requirements for each type of detector.

Since the figure D^* for normalized detectivity includes the aperture area parameter, as well as system bandwidth, it provides a more complete basis of comparison than NEP between detectors. From this column it can be seen that the 10-mm Golay detector (with modifications) affords as high a D^* over the complete 0.5- to 40-micron spectrum as any of the photoconductors when averaged over the 8- to 14-micron range.

As pointed out, in most instances the planet-temperature ranges lie in the 200 to 380 °K range. As seen from Figure 14, the copper-doped germanium detector has a response such that approximately 73 per cent of the emission from a 200 °K source would fall within its passband, and over 94 per cent for a 380 °K source. The D^* exceeds 2×10^9 cm-w from 2 to 30 microns, and as seen it is the most sensitive detector of all at its peak at 20 microns. The zinc-doped germanium type extends the limit for photoconductors the farthest, to about 40 microns. However, it has a sensitivity of only about half that of copper-doped germanium from 2 to 20 microns.

The relatively sharp long-wavelength cutoff noted for all of the photoconductors is largely a function of the activation energy level determined by the impurity material used. Hence, gold doping results

TABLE 3. HIGH-SENSITIVITY DETECTORS SUITABLE FOR OPERATION IN THE 2-40 MICRON RANGE*

Detector Type	NEP-Aver. Over 8-14 μ Band $\Delta f = 1$ cps	$D^* = \dfrac{\sqrt{A}\sqrt{\Delta f}}{NEP}$ Aver. 8-14 μ	τ d	Effective Aperture Area	f_0	Oper. Temp. and Coolant
Syracuse University Ge:Cu for 60° field of view	2.0×10^{-11} w	1.5×10^{10} cm/w	$<1 \mu$ sec	8 mm^2 with integration	900 cps	20 °K liquid H$_2$ or H$_e$ under pressure
SBRC Ge:Cu for 120° field	1.18×10^{-10} w	1.5×10^{10} cm/w	$<1 \mu$ sec	314 mm^2		
Perkin Elmer Ge:Zn (ZIP) for 60° field of view	2.8×10^{-11}	7×10^{9}	0.01μ sec	4		4.2 °K liquid H$_e$
RCA – Ge-Si: Zn, Sb (antimony compensation)	–	8×10^{9}	0.1μ sec	Not known		48 °K Pump on liquid N
10-mm Golay Detector† with optical and photo-tube improvements (Sperry modified)	5×10^{-11}	1.8×10^{10}	2 msec	78 mm^2	10	300 °K – no cooling
Same as high f_0	1.0×10^{-10}	9×10^{9}	–	–	30	
Immersed thermistor Barnes BE-564	9×10^{-11} w	1.05×10^{9} cm/w	7.1 msec	2.55 mm^2	20	

*Only a few of the listed detectors cover this entire range.

†The thermal detectors listed (Golay and Emersed Thermistor) have detectivities D^* and NEP's which are essentially flat over this entire 2-40 micron range, as noted in the text.

in a $\lambda_{co} \approx 10$ microns, copper at 25 to 30 microns, and zinc out to 40 microns. The response of extrinsic photoconductors is a function of the cooling temperature, which must be carefully controlled for stable responsivity in cases where detectors are not operating at liquefying temperatures and/or atmospheric pressure.

Thus, for a sensor required to cover mainly the 2 to 30 micron range, the copper-doped detector operating at 15 to 20 °K would probably provide the best all-around performance, providing the cooling requirements were no real problem. Since for space applications the operation of these devices at near absolute zero temperatures does pose a formidable problem at the present and for the foreseeable state of the art, this problem becomes the primary disadvantage in the use of photoconductors, as will be further discussed.

Copper-doped germanium detectors are relatively simple and economical to fabricate and can be made with large aperture areas and still provide less than 0.1 to 1 msec time constants. Recent studies show that detector areas of over 3 cm^2 can be achieved and at essentially the same D^* range, as shown in Figure 14.

Time Constant. Since the servo actuators for attitude control of the vehicle have bandwidths that are typically limited to a few cycles, the response time constant of the i-r detector should not have to be very fast.

The modulation frequency required while scanning the planet involved is limited by the rotational speeds of the vehicle for the class of attitude sensors that utilize this motion itself for the actual modulation process. This would indicate that i-r modulation frequencies of 10 to 20 cps should be maximum for this method.

Even for an active scanner (i.e., separate internal scanning or modulation drive) the modulation frequency need not be very high, particularly if the useful system bandwidth is in fractions of cycles per second.

Therefore, though the expected limit of modulation frequencies for thermistors and Golay detectors is on the order of 100 cps, this should be more than adequate for most foreseeable usage of i-r detectors in space vehicle attitude sensors.

On the other hand, though the photoconductors can be operated to very high modulation frequencies owing to their submicrosecond time constants, they nevertheless exhibit a $1/f_c$ relation increase in noise figure as modulation frequency f_c is reduced. In most cases this effect starts at and below 200 to 500 cps. For copper-doped germanium detectors, it is claimed that $1/f$ noise is not a factor until the modulation frequency becomes less than 20 cps. But even with this type, operation at or below 10 cps might prove undesirable.

In any case, performance of all of the photoconductors falls off as f_c is reduced far enough, and may render a particular photoconductor unusable if the required modulation is in the 5 to 100 cps range. Therefore, not only is the Golay detector, or thermistor, well suited for this low f_c range, but use of a photoconductor in the same range may be precluded by its $1/f$ noise figure limitations. Or to put it another way, the reduction in D^* capabilities at such frequencies might well render the use of any photoconductor unjustified in lieu of the weight and power penalty paid for the necessary cooling equipment (again assuming the latter is feasible). But in an application where active modulation can be employed, this does not mean to say that an optimum f_c for the detector could not be achieved where such a frequency (probably in the 500 to 1000 cps range) were advantageous in accomplishing the task of scanning and modulating the radiance from around the planet's disc.

It is felt from the foregoing discussion that a detector with capabilities of operating up to 100 cps should prove adequate for most of the attitude-sensor techniques of concern to this report.

Cooling Photoconductors in a Space Vehicle. There are several major problems involved in the operation of a closed-system cryogenic cooler needed for operation of high-performance extrinsic photoconductors in a space environment:

1. The weight required for the cooling hardware and, particularly, compressors and drive are still significant for present and contemplated mechanical schemes using gaseous and liquid coolants.
2. Power requirements are also relatively large for heat-exchanger schemes in either the above category or in utilizing such methods as thermoelectric coolers where no compressor or plumbing is required.
3. The operating time of all presently developed or proposed mechanical coolers appears limited.

One can conclude that without significant advances in these three areas, cooled detectors will not be generally suitable for use in space vehicles. Fortunately, the alternative of using detectors which do not require low temperatures is not unattractive, since thermistors and pneumatic detectors seem well suited by sensitivity, time constant, and power requirements.

Signal Processing

The basic signal processing problem in infrared-sensitive horizon trackers is that of converting the detector output to a vertical-error signal. For this discussion, it is convenient to discuss two different classes of trackers: open-loop instruments, which generate an output signal that is a function of the indicated vertical error directly, and closed-loop instruments, which involve the erection of a platform (possibly the vehicle itself) to the vertical, and indicate the vertical error by measuring an angle between a platform-fixed reference and a vehicle-fixed reference.

Open-loop Instruments — Horizon Sensors. This class of instruments is useful in determining the vertical, but contains no provisions for tracking it. This type apparently is in use on the TIROS satellite to indicate when the spin-stabilized vehicle is oriented in an attitude that permits visual reconnaissance of the earth's cloud cover, but no attempt is made to keep either the vehicle or the horizon sensor field of view pointed at the earth.

In this type of instrument, the indication is obtained by a solution of the equation,

$$\theta_1 - \theta_2 = \Delta\theta$$

which is derived from Figure 15.

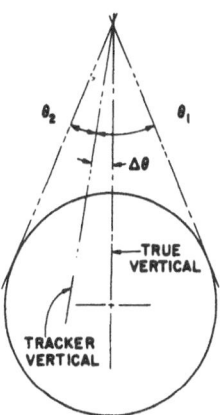

Figure 15. Tracker geometry.

θ_1 and θ_2 can be determined from a measurement of the time at which the field of view intercepts the horizon (in a circular-sweep system), as discussed in [13] and [14]. These times are then compared to an arbitrary time reference in the instrument, producing sample pulses at the scan rate, which are then fed to a holding circuit which holds its value until the following pulse either resets or changes the output. This output is then basically a time measurement which is proportional to the angular vertical error. As such, the measurement is independent of variations in target radiance, detector sensitivity, amplifier gain, and power-supply voltage, since the detector is determining only the time at which the field of view passes across the maximum radiation gradient. This time measurement is useful with circular-sweep systems, and with slight modifications, the same measurement is useful to sector-sweep systems.

Discontinuity trackers, however, seem to be dependent upon an angular measurement between the bisector of the scan-cone apex angle and a reference axis for the vertical-error measurement. This can be made with conventional transducers. The interesting data process taking place in this type of instrument is the actual tracking of the discontinuity. A possible tracking technique using a modulation reticle is described under "Scanning Techniques" in this report. Another scheme, discussed in detail in [14], is based on high-frequency crossing of the discontinuity, as illustrated in Figure 16. The zero-error condition here is defined by pulses that are as wide as the space between pulses. For a positive vertical error the pulses will become narrower than the interpulse space, and for a negative vertical error, the pulses become wider than the interpulse space. With proper shaping of the high-frequency motion, the change in pulse width can be made proportional to the angular error, or at least some predetermined function of the angular error, if the discontinuity is in the range of the high-frequency scan. Nonuniform changes in the pulse width provide a measure of the location of the disc image within the field of view, thus providing a coarse-fine angular-error measurement.

Closed-Loop Instruments — Horizon Trackers. This type of instrument is similar to the horizon sensor discussed in the preceeding section, but goes a step further in that it provides a continuous indication of the present position of the vertical. For a continuous indication either the tracker mount must possess at least two degrees of freedom or the vehicle must be attitude-controlled. The

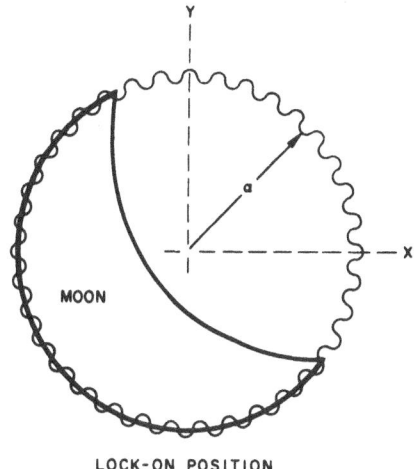

LOCK-ON POSITION

$+\Delta a$ ERROR

$-\Delta a$ ERROR

Δ X ERROR

NO ERROR

Figure 16. Tracking data.

two-degree-of-freedom mounting permits a measurement of the vehicle attitude by determining an angle between the tracker reference
(the vertical) and a vehicle reference; if the instrument has less
than two degrees of freedom, provision must then be made for vehicle
rotation. Thus, one other component is required in the loop—a
source of torque, either to rotate the platform or to rotate the vehicle
to null the error. The major difference is one of torque level; in the
case of platform rotation, gyro-type torques are sufficient.

Other Factors

Other factors that must receive consideration in the design of
horizon-tracking instruments are based on the peculiarities of
satellites and space vehicles. Since the operating lifetime of these
instruments is usually of the order of months, and very probably re-
dundant sensing is not possible, reliability over the lifetime of the

instrument is of prime importance. Other equally as important features of design are low power consumption (5 to 10 w), since operating power must be internally generated or stored, and there must be a minimum number of moving parts. The latter is important not only from the standpoint of reliability, but also from the standpoint that moving parts, in almost all circumstances, are associated with reaction torques on the satellite or space vehicle, and, as such, must be compensated by a stabilizing torque to provide a stable orientation in space. The stabilizing torque, in turn, places extra requirements on the torque-generation system and power supply, thus further penalizing the over-all vehicle system.

Since the system is operable outside the atmosphere, it must be operable in the hard vacuum of space in the presence of cosmic radiations, and possibly must be required to survive a passage through the Van Allen radiation belts. Table 4 is a tabulation of some of the anticipated environmental conditions.

TABLE 4. ESTIMATED ANTICIPATED ENVIRONMENTAL CONDITIONS

Temperature	nonoperating — 0°F to 650°F operating — 0°F to 100°F
Atmospheric pressure	hard vacuum to 15 psi
Radiation dosage	probably 3 to 30 roentgens, minimum
Acoustic noise	138-183 db over 30 cps to 10 kc spectrum
Vibration	40-g load over 5 cps to 2 kc spectrum

The system must also be designed for the obvious requirements of minimum space and minimum weight, and a minimum of protrusions through the skin of the vehicle.

Closed-loop Operation

When a horizon-sensing instrument is operating in a loop to provide platform torque signals, the torque may be supplied to a platform within the vehicle, or to the vehicle itself, so that the vehicle in effect is the stabilized platform. A discussion of vehicle stabilization is beyond the scope of this section, as is the comparison of vehicle stabilization and platform stabilization. It is the purpose here to show an example of how the basic information generated by a horizon sensor might be used to derive a stable vertical, and to disclose any requirements on the sensing instrument that are peculiar to closed-loop operation.

Consider the geometry shown in Figure 15; it is clear from this diagram that θ_1 approaches θ_2 in value, as the tracker vertical approaches the true vertical, or

$$\theta_1 - \theta_2 = \Delta\theta$$

The tracker vertical is defined as a convenient reference on the instrument. These angles, or their difference, must be available in some form from any horizon-sensing instrument. This can be included in a general block diagram, as shown in Figure 17. Fixed time lags are present in the tracker due to at least the time constant of the detector, the inertia of the tracker head, and possibly of the data processor which converts the amplified output signal from the detector into signals proportional to the two angles. The finite time associated with the scanning process can be included in the data-processing time, and may be a time delay rather than a linear time lag. If this delay is appreciable compared to the system time lags, the loop must be analyzed, using sampled-data techniques as discussed, for instance, in [18]. If some irregular surface like the planet-space interface or an atmospheric horizon is chosen as *the* horizon, the stabilization networks shown can have a time constant of the order of hours or days to provide sufficient smoothing for a highly accurate vertical in the presence of horizon irregularities. If the required accuracy is of the order of one degree or more, the smoothing times can be much shorter. The torquer is a source of power necessary to drive the tracker vertical to the true vertical, which is sensed when $\Delta\theta$, the difference angle, goes to zero. This device will have the usual dynamic characteristic of one integration with an associated time lag.

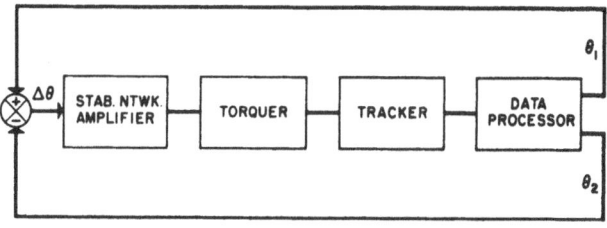

Figure 17. Block diagram.

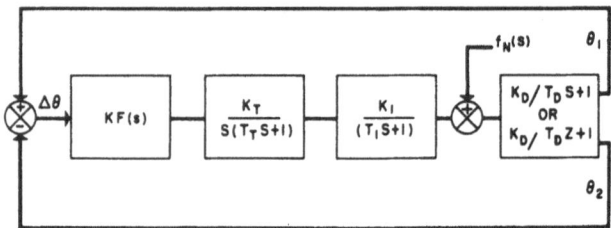

Figure 18.   Loop dynamics.

The dynamics of the loop are illustrated in Figure 18, where $F(s)$ is the smoothing-filter transfer function and $f_N(s)$ is the frequency characteristic of the surface irregularity. The $f_N(s)$ characteristic is essentially a noise input to the loop, and can be treated analytically as such.

The noise input characteristics can be computed from the power spectral density. From a physical map, the autocorrelation function of altitude versus ground distance can be computed by reading altitude at small intervals along a great circle. As a typical simple great circle, the equator was chosen, and the computation performed. Figure 19 shows the autocorrelation function and a simple exponential fit of the form,

$$f(t) = K + C\, e^{-T/T_c}$$

where

$$K = 1.7 \times 10^4$$
$$C = 17.5 \times 10^4$$

A better fit could probably be obtained by using the sum of two or more exponential functions, but the accuracy of the data and the purposes for which the data are intended does not warrant the extra complication at this time. The parameter T_c was taken as 3.5 for the fit shown. T_c is actually a correlation time measured in degrees; this must be converted to time units for convenience and later use. For a satellite, this can be done by

$$T_t = \frac{T}{360} \times 3.5$$

where T is the satellite's orbital period in hours, and

$$\omega_t = \frac{360}{3.5T}$$

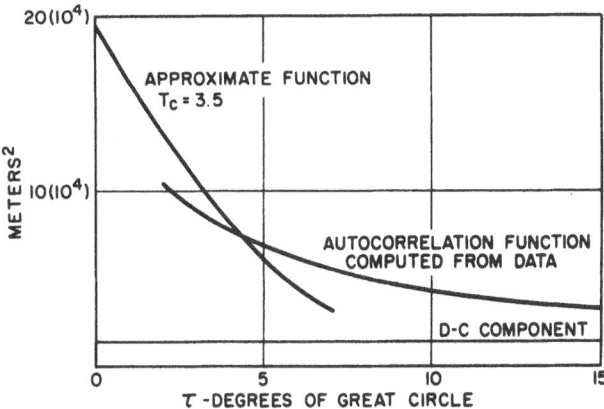

Figure 19. Autocorrelation function of earth equatorial surface
irregularities.

In the time domain, the nonconstant part of the autocorrelation
function may be represented by

$$\theta(t) = C\,e^{-360/3.5\ \tau/T}$$

By the Wiener-Kitchine relationship relating the autocorrelation
function to the power spectrum,

$$W(f) = 4\int_0^\infty \phi(t)\cos \omega\tau\, d\tau$$

$$\Phi(w) = \frac{4020 \times 10^4}{\left(\dfrac{360}{3.5T} + j\omega\right)\left(\dfrac{360}{3.5T} - j\omega\right)}$$

The coefficient retains the dimensions of square meters; this
may be converted to square radians by dividing the coefficient by
the square of the sight-line range to the horizon, thus giving the
power spectrum in a form convenient for analysis.

REFERENCES

1. R.E. Roberson, "Optical Determination of Orientation and Position Near a Planet," paper presented at the ARS Semi-Annual
   meeting, Los Angeles, California (June 1958).
2. W. Chauvenet, A Manual of Practical and Spherical Astronomy,
   New York, Dover Publications, 1960.
3. Handbook of Geophysics, Geophysics Research Directorate,
   U.S. Air Force Cambridge Research.

4. *Exploration of the Moon, the Planets and Interplanetary Space,* Report No. 30-1 JPL CIT, April 1959, NASA Contr. No. NASW-6.
5. K.A. Ehricke, *Space Flight,* Vol. 1, "Environment and Celestial Mechanics," Princeton, N.J., Van Nostrand, 1960.
6. Radiation Lab. Series, Vol. XIII, *Propagation of Short Wave Radiation,* New York, McGraw-Hill.
7. H.C. Van de Hulst, "Scattering in the Atmospheres of the Earth and Planets" in *The Atmospheres of the Earth and Planets,* ed. G.P. Kuiper, Chicago, University of Chicago Press, 1952.
8. E. McCartney, "Horizon Tracker for Atmospheric Reentry" AAS Preprint No. 58-49 presented at the 5th Annual Meeting, Washington, D.C., December 1958.
9. L. Larmore, "Range Equation for Passive Infrared Devices," *Proceedings of the IRE,* **47**: 9, (September 1959).
10. R.H. Genoud, "Infrared Search System Range Performance," *Proceedings of the IRE,* **47**: 9 (September 1959).
11. R.M. Scott, "Optics for Infrared Systems," *Proceedings of the IRE,* **47**: 9 (September 1959).
12. M.E. Lasser, P.H. Cholet, and R.B. Emmons, "Electronic Scanning for Infrared Imaging," *Proceedings of the IRE,* **47**: 12 (December 1959).
13. M.H. Arck, and M.M. Merlen, "Horizon Sensors for Vertical Stabilization of Satellites and Space Vehicles," paper presented at the IAS National specialists meeting on Guidance of Aerospace Vehicles, Boston, Mass. (May 25-27, 1960).
14. R. H. Grube, "Terminal Guidance for Lunar and Planetary Probes," presented at the IAS National Specialists Meeting on Guidance of Aerospace Vehicles, Boston, Mass. (May 25-27, 1960).
15. P. Elias, D.S. Grey, and D.Z. Robinson, "Fourier Treatment of Optical Processes," *Journal of the Optical Society of America,* **42**: 2 (February 1952).
16. G.F. Aroyan, "The Techniques of Spatial Filtering," *Proceedings of the IRE,* **47**: 9 (September 1959).
17. R.E. Roberson, "Methods of Attitude Sensing," *Proceedings of Manned Space Stations Symposium at Los Angeles, Calif.* (April 20-22, 1960); Sponsored by the Institute of the Aeronautical Sciences in Cooperation with NASA and The RAND Corp.

SOFT LUNAR LANDING GUIDANCE SENSOR

A. BARABUSH

Missile Systems Division, Raytheon Company,
New Bedford, Massachusetts

In the near future United States space vehicles, such as the Centaur, will be soft landing 300 to 730 lb payloads on the moon. By the early 1970's, the Nova vehicle will be landing payloads of 20,000 lb — enough payload to permit a man to go to the moon and come back. The lack of an appreciable atmosphere on the moon rules out the possibility that these space vehicles could make soft landings based on the use of aerodynamical devices (parachutes, lifting surfaces). In lieu of this, these space vehicles will carry astronautical reaction-type devices (retrorockets, etc.) to be used to effect the soft landing. As a corollary, the moon vehicles will carry a lunar landing sensor capable of providing the necessary intelligence for control of this energy. An initial design of such a soft lunar landing guidance sensor is the subject of this paper [1].

The initial lunar vehicle flight path will be modified during flight by midcourse corrections to improve guidance accuracy [2, 3]. These corrections will be based on information from radar located on the earth or self-contained optical-inertial systems. In the vicinity of the moon the landing vehicle will be rotated so that the retrorockets face the moon (see Figure 1), the verticality of the landing vehicle being controlled by an i-r/optical horizon scanner. The last 10 minutes of the flight will then be controlled by the landing sensor.

A natural question at this point might well be: Why a separate landing sensor [4, 5]? Why not use the optical-inertial system or the i-r/optical horizon scanner to control the landing? Basically the reason is this: The use of information obtained several lunar radii away from the moon to control the landing depends too heavily on system calibration (precise rocket thrust, precise thrust timing) — it is open-loop control. If the moon surface altitude were estimated

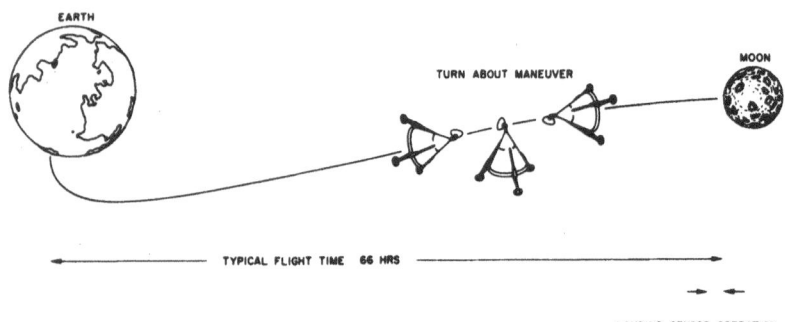

Figure 1. Lunar flight trajectory.

incorrectly by only 100 ft, open-loop control would result in a
velocity error of 22 miles/hour, which is more than can be allowed.

$$v_f^2 = v_0^2 + 2gs$$

where:

v_f = impact velocity
v_0 = zero velocity at predicted altitude
g = acceleration due to moon's gravity = (0.16)(32 ft/sec)
s = 100 ft misprediction

$$v_f^2 = v_0^2 + 2(0.16)(32)(100) \doteq 22 \text{ miles/hour}$$

Coherent-type radar is the only present self-contained means that
provides continuous relative velocity (vector) information. Radar
information is capable of providing closed-loop control of the
vehicle landing until final touchdown is accomplished.

The landing sensor must provide three quantities: the vehicle
altitude, its vertical velocity (vector), and its lateral velocity
(vector). With these quantities known, the lunar landing can be
controlled. The required lunar landing sensor performance is given
in Table 1. The table is not based on any specific systems, but
rather typifies the values expected. The initial operating parameters
are the conditions at the time the sensor is asked to assume command of the landing.

TABLE 1

Initial operating parameters:	System A	System B
Altitude	60,000 + 10,000 ft	30,000 ± 5,000 ft
Vertical velocity	−7000 ± 700 mi/hr	0 ± 200 mi/hr
Lateral velocity	0 ± 700 mi/hr	0 ± 200 mi/hr
Required landing sensor performance:		
Altitude accuracy	± 1%	± 5%
Velocity accuracy	± 7 mi/hr*	± 1 - 2 mi/hr*

*In vicinity of zero velocity.

LUNAR LANDING RADAR DESIGN

Basic Radar Phenomenon

Lunar landing radars will make use of the scattering properties of natural surfaces. A radar beam striking a smooth polished surface at an oblique angle will be reflected away; a radar beam striking a rough surface will be scattered, and some of the scattered energy will be returned in the direction of the transmitter (Figure 2). Surfaces found in nature are typically rough (on a scale comparable with the radar energy wavelength). The energy returning to the radar will be frequency-shifted (Doppler) by the relative velocity between the transmitter and the individual scatterer. This phenomenon allows the use of *oblique* radar-energy impingement for measurement of velocity data, a necessary requirement for measuring lateral velocities (Figure 3). As a consequence, the information from three non-coplanar radar beams aimed at the moon's surface, together with a vertical reference system, can completely determine the lunar vehicle velocity (vector).

Type of Radar [6]

The radar could be straightforwardly instrumented in many cases with a CW system (transmitter continuously sending and receiver continuously receiving the scattered Doppler shifted signal). For the equipment to be described, aperture limitations prevented the use of a two-aperture-per-radar-beam system (that is, a straight CW system requires a separate transmit and receive antenna in order to achieve the necessary isolation between transmitter and receiver).

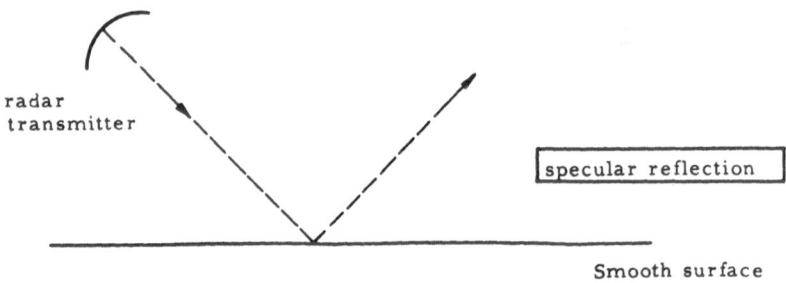

specular reflection

Smooth surface

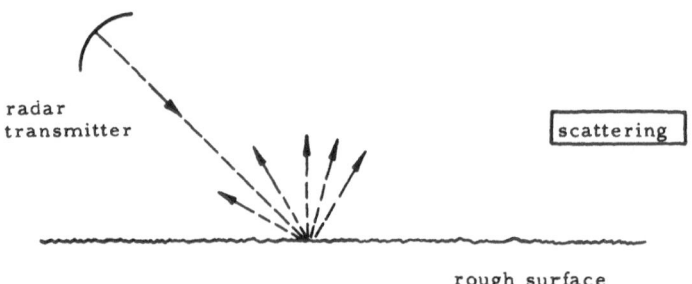

scattering

rough surface

Figure 2. Scattering phenomenon.

In the designed system the isolation is achieved by FM-CW techniques.

FM-CW Radar [7, 8]

The design complexity of an FM-CW system is only slightly greater than that of a CW system. The basic advantage of an FM-CW system is that with a single antenna it provides the necessary transmitter-receiver isolation. This can be shown by the following mathematical analysis: Let e_1 equal the transmitted FM-CW signal and e_2 the return signal — still the same FM signal, but with a propagation time delay of T and a shift of all the FM frequencies by the Doppler frequency:

$$e_1 = A \sin \left(Wt + \frac{\Delta F}{\mu} \sin \mu t \right)$$

$$e_2 = B \sin \left[W_1 (t - T) + \frac{\Delta F}{\mu} \sin \mu (t - T) \right]$$

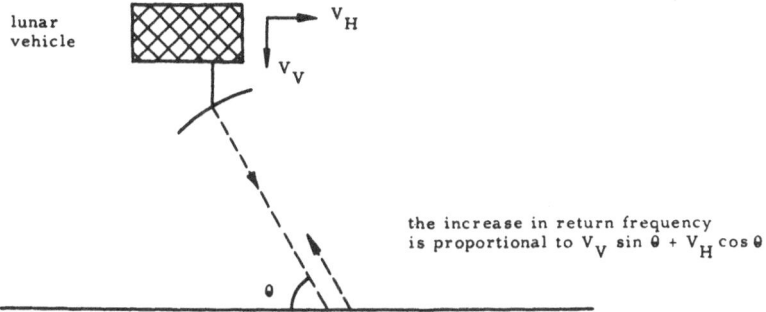

lunar vehicle

V_H

V_V

the increase in return frequency
is proportional to $V_V \sin \theta + V_H \cos \theta$

θ

Moon's surface

V_V = vertical velocity lunar vehicle
V_H = lateral velocity lunar vehicle

Figure 3. Doppler shift.

where:

W = carrier frequency (rad)
μ = FM frequency (rad/sec)
ΔF = frequency deviation (rad)
W_1 = received frequency (rad/sec)
T = delay time (twice range to ground/speed of light)

the difference term in the product e_1 times e_2 is

$$C \cos \left[\left(Wt + \frac{\Delta F}{\mu} \sin \mu t \right) - W_1 (t - T) - \frac{\Delta F}{\mu} \sin \mu (t - T) \right]$$

$$= C \cos \left[Wt - W_1 (t - T) + \frac{\Delta F}{\mu} \left\{ \sin \mu t - \sin \mu (t - T) \right\} \right]$$

leading eventually to

$$C \cos \left[Wt - W_1 (t - T) + 2 \frac{\Delta F}{\mu} \sin \frac{\mu T}{2} \left\{ \cos \left(\mu t - \frac{\mu T}{2} \right) \right\} \right]$$

An examination of the last equation reveals that if T, the propagation time delay, is zero and $W = W_1$ (as would be the case for energy spilling over from transmitter to receiver), there is no AC receiver signal. A signal received with an appreciable T, as compared with the modulation period, will cause a frequency-modulated AC return after mixing. The equation shows that the returned signal modulation index varies with T, and that the signal will disappear at altitudes causing propagation delays equal to the FM period. However, in actual practice the variation in slant range to the ground across the antenna pattern smears the time delay. Nulls in the received signal do not occur except at very low altitudes — one, two, or three times the distance corresponding to the FM period. For a 9 mc frequency-modulated system, this corresponds to altitudes of 55, 110, and 165 ft. This effect is eliminated in the equipment herein by switching to straight CW operation at the extremely low altitudes above. The signal return at this point is so large that isolation between transmitter and receiver is not a problem.

LUNAR LANDING RADAR

Equipment Specifications

The specifications of the lunar landing radar are given in Table 2. An artist's conception of the lunar-radar lunar vehicle in operation is shown in Figure 4.

TABLE 2. LUNAR LANDING RADAR PARAMETERS

Velocity Measurement Section	
Radar Type	FM/CW
Number of Beams	3
Transmission Frequency	35 kmc (K_a band)
Transmitted Power	3 w CW
Antenna Gain	32 db
Composite Beam Width	4°
Altimeter Section	
Radar Type	FM/CW
Number of Beams	1 (uses one of vel. unit antennas, frequency-duplexed)
Transmission Frequency	32 kmc
Transmitted Power	1 w CW
Complete Lunar Sensor (less primary power)	
Volume (including antennas)	2 cu ft
Volume (electronics package)	1 cu ft
Aperture Area	0.37 sq ft
Power Required	300 w
Weight	50 lb
Accuracy Velocity Information	± 1.5%
Accuracy Altitude Information	± 5%
Maximum Operation Altitude	60,000 ft
Minimum S/N Required	± 5 db
System S/N @ 60,000 ft	± 15 db

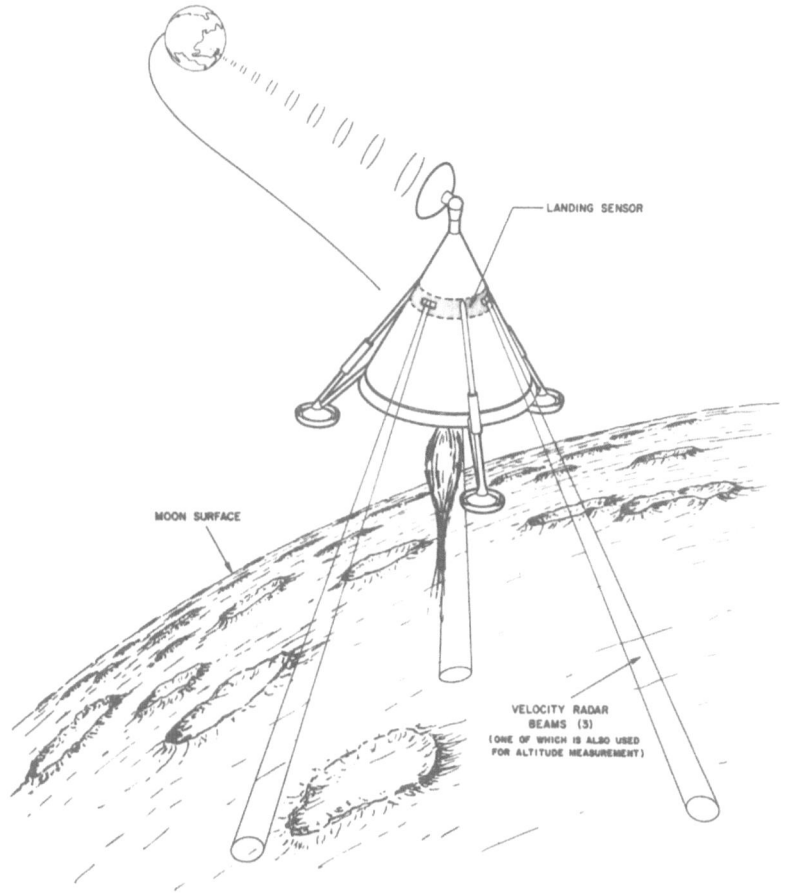

Figure 4. Lunar landing system configuration.

Block Diagram

The block diagram of the conceived lunar landing sensor is shown in Figure 5. The transmitter operates at K_a band (35 kmc) and employs a set of three antennas generating three symmetrically placed beams. Transmitter-receiver isolation is obtained by the FM frequency-duplexing previously described. A 9 mc signal is used to place FM sidebands at intervals of 9 mc about the 35 kmc carrier, with a modulation index of three. The transmitter power of 3 w is split into three separate channels and transmitted via the duplexers. Upon reception the return signal is mixed with the local oscillator

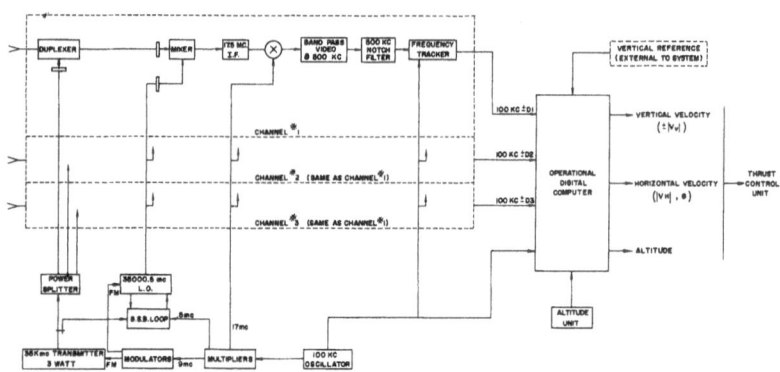

Figure 5. Lunar landing sensor — block diagram.

signal to obtain the various FM sidebands. An i-f amplifier set at one of the sidebands (the second sideband at 17.5 mc) extracts the desired frequency-shifted spectrum. Figure 6 shows the system frequency spectrum. The top portion shows the frequency spectrum of the frequency-modulated transmitted signal. The received signal is shown in the lower half. The Doppler frequency shift causes a shift in frequency of all of the components of the received spectrum.

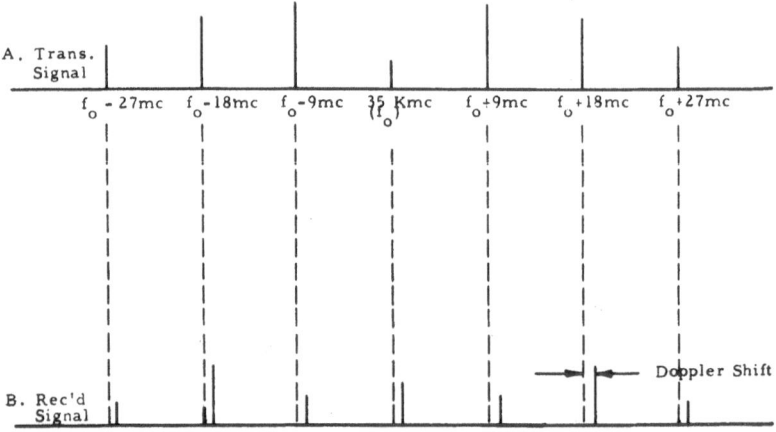

Figure 6. Transmitted and received frequency spectrum.

The local oscillator is side-stepped from the transmitter frequency by 500 kc. If the transmitted frequency were used for the local oscillator signal the spectrum as seen in Figure 6 would be folded about f_0. In this case the Doppler signal associated with the lower 18-mc FM sideband and the Doppler signal associated with the upper 18-mc FM sideband would appear about 18 mc i-f center-separated by twice the Doppler. With this system the speed could be determined, but not the vector, i.e., equal velocities in either direction would produce the same spectrum at the 18-mc i-f. The side-stepped local oscillator, however, separates the upper and lower FM sideband Doppler by twice the Doppler frequency plus 1 mc. The Doppler return is thus separated into a signal which has unambiguous velocity vector information associated with it. A 17.5 megacycle signal represents zero velocity, as shown on the block diagram. Vehicle velocities of 200 miles an hour, causing Doppler frequencies of 14 kc, would appear, for positive velocities, 14 kc above the intermediate frequency and, for negative velocities, 14 kc below the intermediate frequency.

The Doppler signal is then passed to a 500-kc-centered video (second i-f) amplifier for additional amplification. A notch filter is inserted in the video amplifier at the exact center frequency, so as to reduce noise signal caused by the noise characteristics of the transmitting klystron. Transmitter noise places some noise energy in the vicinity of the zero Doppler frequency band. The notch compensates the effect of this noise. The signal then passes to the frequency tracker. The received signal up to this point has characteristics similar to white noise passed through a bandpass filter. For the system beam width of 4 degrees, the band width of the Doppler signal is roughly 20 per cent of the Doppler frequency. In order to (1) improve the signal-to-noise ratio and (2) determine the center frequency of the signal, the frequency tracker is employed actually to determine the power center of gravity of the spectrum. The frequency tracker is shown in more detail in Figure 7. Search functions are provided in the frequency tracker to acquire the signal initially.

The signal from the frequency tracker (100 kc ± the Doppler frequency) is then passed to the operational digital computer. This Doppler information, together with the Doppler information determined by the other two channels, plus vertical information, provides all of the necessary information for the determination of the vertical velocity and lateral velocity of the vehicle. This information is then passed to the thrust control unit. The velocity portion of the computer is shown in Figure 8.

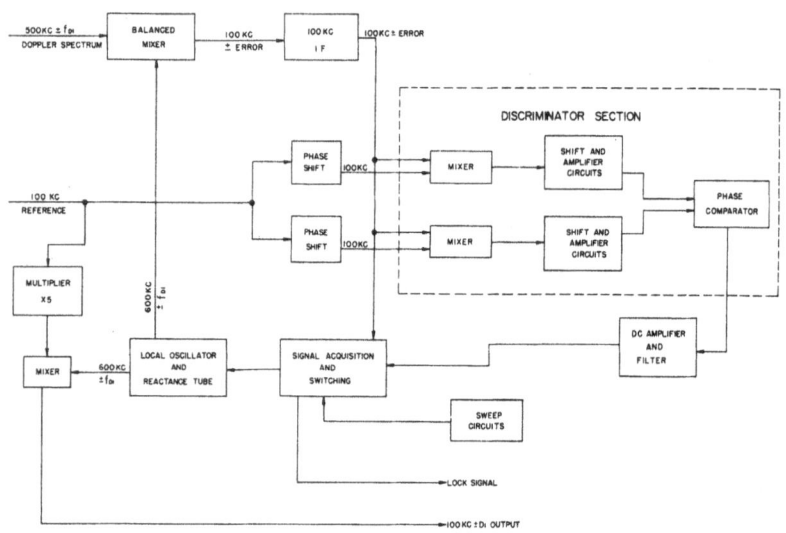

Figure 7. Frequency tracker — block diagram.

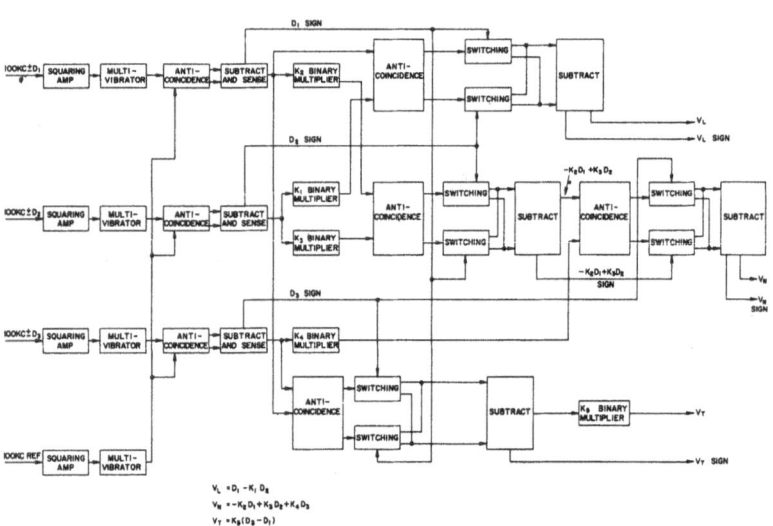

Figure 8. Velocity computer — block diagram.

On the block diagram it will be noted that the altimeter function is provided by a separate unit. The question always arises in the design of a lunar landing radar as to whether to combine the altimeter function into the velocity-measuring transmitter-receiver or to provide a separate transmitter-receiver. Because of the complex modulation of the FM-CW velocity system it was deemed advisable in this system to provide a separate altimeter transmitter-receiver (particularly since the altimeter operation is performed with high modulation index FM-CW, as contrasted with the low FM modulation index velocity system). The altimeter makes use of one of the velocity unit antennas on a frequency-duplexed basis. The altimeter operation is based on the standard FM-CW triangular modulation techniques, where high modulation index-wide frequency deviation techniques are employed [9].

SPECIFIC DESIGN POINTS

Several important points deserve more specific attention than allotted in the block diagram discussion above; these are commented on below.

Weight and Reliability

It may be rather shocking to see such mundane subjects as weight and reliability heading a list of specific points to consider in the design of a lunar landing sensor. For those engaged in the field, however, both are extremely serious considerations. The story of the Atlas-Able shots indicates the absolute necessity for high reliability, and since it takes approximately 1,000 lb of vehicle on the earth to land 1 lb of equipment on the moon, the importance of the latter item is self-evident.

Accuracy

Navigation equipments, presently in aircraft use, are capable of achieving velocity accuracies of 0.1 per cent and 1 per cent of the altitude. However, the lunar vehicle system described herein is directed toward a velocity-determination accuracy of 1.5 per cent and an altitude-determination accuracy of 5 per cent. Since the entire system is in effect a zero nulling servo loop, the percentage accuracy need not be too high. Extreme accuracy is only necessary at the low velocities and low altitudes near touchdown. At this point bias errors become the dominant factor, rather than the percentage error. For the present equipment, the maximum bias error is about 1 mile per hour in velocity and 25 ft in altitude.

Radar Cross Section and Sensitivity [10]

A lunar landing radar must be designed with a proper consideration of the probable reflecting characteristics of the lunar surface. Theoreticians, under the assumption that the moon's surface is an isotropic scatterer, have predicted the path loss to be expected by a radio wave which is back-scattered from the surface. Experimental evidence across a wide radio spectrum, 150 mc to 3 kmc, indicates that the actual reflectivity is 6 db or more greater than indicated by the mathematical model. This seems to indicate that some specular reflection occurs. Added credence for this belief is furnished by comparing the theoretical and actual pulse shapes of the echoes. As illustrated in Figure 9 the principal echo power in the actual return is confined to a region which is small compared with the moon's radius, while the theoretical echo has significant return from a larger fraction of the hemisphere.

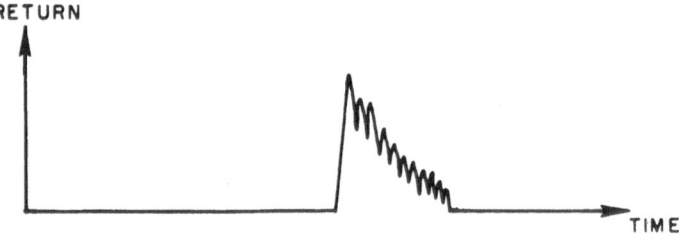

(a)   FORM OF PULSE RETURN IF MOON WERE ISOTROPIC

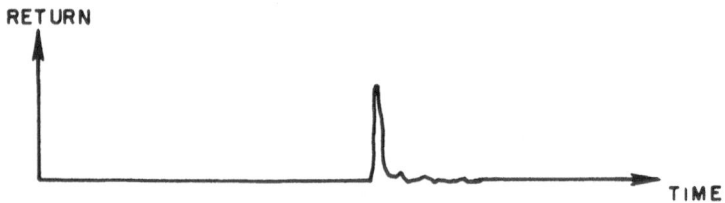

(b) FORM OF EXPERIMENTAL PULSE RETURN FROM MOON

Figure 9. Radar signal forms.

These results must be regarded as preliminary, and cognizance must be taken of the contrasting conclusions drawn by recognized experts. Even if unchallenged and detailed radar characteristics of the moon were known, they would be of only limited value to the lunar landing sensor. This results from the uniqueness of the moon as a radar target. Conventional radar targets can be categorized as point targets (aircraft), area targets (surfaces), and volume targets (rain). In the last two instances it is assumed that targets cover the entire beam width and the reflectivity is measured in terms of radar cross section per unit area or volume. As illuminated from the earth, the moon is much larger than a point source and yet fails appreciably to fill the antenna beam width utilized to date. Consequently, if conventional radar equations are used, the lunar radar cross section depends on the particular radar parameters and the earth-moon range in an abstruse way.

It is interesting to speculate which of the variety of surfaces encountered on earth most nearly resembles a lunar "sea" in terms of radar reflectivity. Reflectivity curves typically encountered on the earth are illustrated in Figure 10. At the present time a dry desert appears to be a likely choice. Based on data by Grant and Yaplee of NRL, dry desert appears to be somewhere between the bottom two curves.

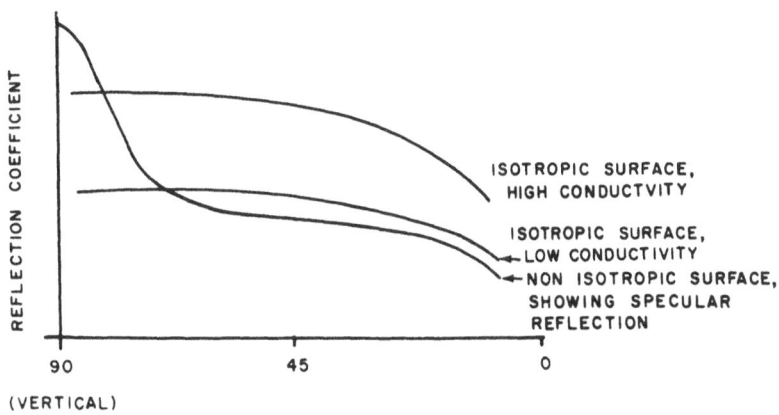

Figure 10. Typical reflectivity curves.

It will be noticed that the curve of Figure 10 does not include any numerical values of reflection coefficient. Various authorities and experts have estimated the expected moon reflection coefficient at the antenna beam incidence angle of 70°. Estimates vary from a −17 db to −30 db reflection coefficient. Minus 20 db is a reasonable estimate and is accepted by some authorities.

Miscellaneous

In addition to the above major effects several other unique items have to be considered in the design of a lunar landing system.

One item to be considered is that the exhaust plume of the retro-rocket contains highly ionized gases capable in some cases of attenuating r-f energy. Solid fuel, metal additives have been known to cause 1 db attenuation per 1 per cent additive for radar energy passing through the plume. In the design presented the beam angles are at 20 degrees from the vertical and are not bothered by these effects (Figure 4). If side jets are used for control of the lateral velocity, the jets should be interspaced between the radar beams.

The 35 kmc lunar radar should not be bothered by any ionized sheaths around the moon. The moon's atmosphere (10^{-13} atm at most) is too sparse to cause any significant attenuation of the milli-metric waves. An item that deserves consideration is the possibility that the retrorocket will impinge on what some experts claim is a moon surface composed of a mile-deep covering of low-density dust. Two serious problems could arise: (1) the lunar instrument package may dig in and bury itself, and (2) the lunar landing radar may indicate erroneous information from radar energy scattered by the great mass of dust particles. Most experts, however, believe the moon's surface does not have more than a few centimeters of dust covering.

The lunar landing equipment electronics could be used to transmit information back to the earth after landing. In this case it would be advisable to reduce the radar's operating frequency to below 10 kmc so as to permit transmission through the earth's atmosphere in the dual mode [11]. Finally, radars for operation in space must be designed to withstand the sterilization process contemplated for all space equipments. One method of sterilization involves "roasting" the equipment at 125 °C for half a day. The effects of this processing should be considered in the design of the equipment.

REFERENCES

1. H. Newell, "Capabilities for Space Research," in *Exploration of Space*, ed. R. Jastrow, New York, Macmillan, 1960.
2. Technical Reports of Jet Propulsion Lab., Pasadena, Calif., California Institute of Technology; see, for example, T.R. #32-28.
3. H. Bement, "Lunar Guidance," *Astronautics* (September 1960).
4. R. Sohn, "Lateral Speed Indicator," *ARS Journal* (April 1960).
5. C. Hendrix, "Proposal for a Simple System for Achieving Soft Landing of a Rocket Vehicle," NOTS, China Lake, California, Technical Publication 2495 (unclassified).
6. Walter Fried, "Doppler Radars for Guidance," *ARS Journal* (December 1959).
7. K. C. M. Glegg, "Continuous Wave Doppler Radar with Low Noise," Meeting Notes of Aeronautical Electronics Conference, Dayton (1958).
8. R. K. Brown, "A Lightweight and Self-Contained Airborne Navigation System," *Proceedings of the IRE* (May 1959).
9. D. Luck, *FM Radar*, New York, McGraw Hill, 1949.
10. T. B. A. Senior and K. M. Siegel, "A Theory of Radar Scattering by the Moon," *Journal of Research*, **64D**:3 (May-June 1960).
11. George Haydon, "Optimum Frequencies for Outer Space Communications," *Journal of Research*, **64D**:2 (March-April 1960).

PANEL DISCUSSION: "IS THERE A NEED FOR A MANNED SPACE LABORATORY?"

Moderator: Alfred M. Mayo, Assistant Director for Bioengineering Office of Life Science Programs, National Aeronautics and Space Administration

Panel Members:

Dr. Jesse Orlansky, Member of the Technical Staff, Institute for Defense Analyses, Washington, D.C.

Prof. Ross Adey, Professor of Physiology and Anatomy, University of California, Berkeley, Calif.

Dr. Ashton Graybiel, Director of Research, Naval School of Aviation Medicine, Pensacola, Florida.

Dr. Fred Singer, Physics Department, University of Maryland, College Park, Maryland.

INTRODUCTORY STATEMENTS

Mr. Mayo: The usefulness, or lack of usefulness, of a directly linked human being in space-exploration systems has been the subject of continuing controversy. On the one hand, many outstanding authorities have centered their point of view on the weaknesses of the human being. The fact that he can tolerate but small changes in environment, that his reaction time is slow, and that at times he makes different decisions on the basis of what appear to be the same facts, constitutes perhaps the core of these objections. The proponents of manned control, on the other hand, have based their arguments on the flexibility of man's mental and physical performance, his complex observational capabilities, and his unique ability to arrive at decisions from a complex mixture of stored and immediately observed data. Both sides have presented compelling arguments for automatic versus manned systems, and both can cite examples of success in systems utilizing their philosophy and failures in systems based on the opposite philosophy. Perhaps some reflection on a few noncontroversial points may help in defining a logical approach utilizing the strong points of both the proponents of automation and of manned flight.

Two of these points are:

1. Knowledge to be of importance to human beings must be linked with the human brain.

2. Distance per Coulomb's Law, intervening atmospheres, and particulate matter, such as dust particles, meteoroids, and planets to varying degrees, attenuate the fundamental availability of information transmitted through space.

The potential resolution of information per unit of observation time increases greatly as intervening obstacles and distance are reduced. In light of these facts it would appear that both remotely controlled and directly manned space vehicle systems will be required. The location of the man in the system would be determined by the amount of detail required in the information and the available observation time. It would also be a function of the technological utility of automatic devices capable of being produced at the time the vehicles are defined.

With these considerations in mind, our panel will discuss some new data available and some of the physical science and life science requirements bearing on the need for a manned orbital laboratory. Dr. Jesse Orlansky will discuss human performance. Professor Ross Adey will discuss some of the problem aspects related to human and animal physiology and anatomy. Dr. Ashton Graybiel will provide some important new data on human performance and adaptation to stress during rotation. Dr. Fred Singer will integrate these discussions with some of the applicable physical principles.

Dr. Orlansky: Is a manned space laboratory needed in order to study human performance? Only a single answer is possible, and that is yes. Man has not yet been in space. Our knowledge of his performance on earth does not permit us to predict exactly how well he can perform in space. Therefore, we could use a manned space laboratory to determine the functions which man can perform there. Such data are required for purposes of system design.

Next, I want to identify those aspects of human performance which can be explored in such a facility. These include the ability of man to perform while exposed to extreme atmospheric conditions such as radiation, low-oxygen partial pressure, high and low temperature, carbon monoxide and other atmospheric contaminants. There is no firm basis for predicting man's behavior under prolonged exposure to low g or complete weightlessness. In making a survey, I found that some "experts" believe weightlessness will not be a problem for man (of course, this simplifies the problem of vehicle design); others say it will. The fact is that no one knows how to answer this question, which is why we need a space laboratory.

There are similar problems with vibration, noise, sensory deprivation, and the atypical day-night cycle. We also know that man in space will have to live on a restricted diet, but do not know whether this will affect his performance.

These are some of the physiological stresses to which man would be exposed in space. Now, let us come to some psychological, as opposed to purely physiological, stresses.

Man will be alone in space and such isolation may affect his performance. So may the number and type of individuals in the crew. For example, is it more desirable to be alone or with a team of men? There are also such matters as confinement — that is, restriction to a limited amount of space — and the degree of comfort (as opposed to mandatory environmental conditions) which may affect man's performance.

Finally, some scientists conjecture that intelligent organisms may exist in space. Meeting such organisms could well be a novel problem for our space pilots, who must also adjust to earth on their return.

Now, I hope you agree with me that there are many unknown questions concerning man's effectiveness in space. A space laboratory would be a most useful facility for increasing our knowledge of what man can realistically do in a space environment.

Professor Adey: As a point of departure in this discussion, we must, I assume, first justify the need to have a manned space station. In other words, can the unmanned monitors provide all the necessary information? Can man survive in space? Can he perform and contribute usefully in the space environment? Are there special experiments in human physiology that can only be tested in space, and can be justified as to their scope and expense and even the expendability of man himself?

Now if one follows the prevailing engineering opinion, there are two points of view. One is that manned space flight is unnecessary and will remain so for many years, and if it were not for the fact that this opinion is uttered in such very high places by highly esteemed individuals, one might suspect that it was uttered in arrogant ignorance of the precise situation. But one can take some comfort from the remarks of those like Dr. Masterson, the consultant meteorologist on the TIROS weather satellite project, who recently remarked that, "We too often forget that TIROS is only an observer sending information to human minds for human interpretation. TIROS cannot predict the weather. It takes human intelligence to do that, and this intelligence is a result of the efforts of all of us connected with the project.'' There is indeed a place for man's interpretative capacities in the space environment, to use this tremendous plasticity that he has, to assess data on the spot, and to make decisions that can scarcely be made by any machine that we currently comprehend.

The second engineering opinion that one hears is that when the bird is safe, man will ride along. In other words, man will go merely as a passive passenger. If we look a little further into the picture, he is usually reduced to this role by the lack of backup knowledge of his performance capability in the space environment. But before we go into the question of the lack of that knowledge, I think we can categorically state that there is no mechanical computer that can pack into 150 pounds the capacity that man has to make decisions and to manipulate situations which are very often unforeseen. Quite apart from this facet of the general problem, there is the aspect of man's behavior, inherent in the fact that he is a primate ape, if you like, and cannot reject 50 million years of primate curiosity and primate behavior which will always make him seek new frontiers. Thus one might facetiously remark that when Columbus wanted to sail from Europe to look for North America there were inevitably those who regarded this as a quite improper proposal — a plan destined to failure and quite useless. Presumably, if we go even more facetiously into archeological history, as the lungfish took its first toddling strides from the sea to the mud and on to dry land, there must have been more conservative elements in the lungfish community that resented this step and felt that it was ill-advised. For my part, I think that this step that man proposes to try — even if he fails at least for the time being to go into space — is one comparable in significance to that which took the lungfish from the sea to a land-living environment some 600 million years ago. If we accept the importance and significance of this challenge, we must define the areas which we know to be likely to limit man's performance in prolonged space flight.

This situation is inherently different from sending a man on a ballistic flight downrange or on a few orbits of the earth. For instance, the effect of prolonged weightlessness on the sleep-wakefulness cycles of mammals is quite unknown. But we do have some quite interesting collateral data. For instance, the porpoises and whales are virtually unsleeping, and this state of sleeplessness is associated with an inability to breathe spontaneously. It is true that the porpoises never seem to sleep, and this we suspect may be due to the fact that they are virtually weightless and are deprived of the many barrages of impulses that come from joints and muscles to those of us not living in a weightless environment. This aspect of spontaneous breathing, or the lack of it, in these animals may be quite surprising and even alarming. If an anesthetic is given to a porpoise or a whale to the point where it loses consciousness, it dies immediately. In other words, the animal no longer breathes while unconscious, and we do not yet know whether man will be able to go to sleep and still breathe in the weightless state.

There are well documented cardiovascular effects which occur at least immediately following attainment of weightlessness. We do not know what the long-term effects will be. Secondly, there are the questions of acceleration, noise, and vibration and the effects of irradiation, specifically the effects of irradiation on the brain. The solar flares which come from sun spots from time to time are in themselves a great hazard, and at the moment, for instance, it would take something like seven inches of graphite around a Mercury capsule to afford even average or nominal protection. For the worst solar flare ever recorded, seven feet of graphite around the capsule would be needed, and this would raise its weight to some hundreds of tons. We now come to the question of a combination of stresses. This is something about which we know very little. What happens if we combine weightlessness, irradiation, altered gaseous environments, etc., and we expose a man to the complex for a very long time? The final and pivotal point in man's ability to perform would be the way his brain continues to function in this type of situation.

I think that if we were to sum up at this point, it would be necessary to accept the view that we need an excellent biological and animal backup program as a continuing aspect of space studies, including the feasibility of survival from ground studies. I think it is interesting at this stage to look at the opinions of Russian investigators. Voronin recently said that further experiments will be carried out with monkeys, and this will be followed by observations of man in conditions near those to be met in space. Finally, on the basis of data on man's organism, corrections would be made on the basis of the results of experiments on animals. He further pointed out that in all this there is, in fact, one more difficulty, namely, to ensure the normal functioning of the higher register system, the mental activity of the astronaut.

In our own laboratories at U.C.L.A., under the sponsorship of the United States Air Force Office of Scientific Research, and particularly with the generous assistance of General Flickinger, we have recently attempted to study some aspects of brain function in conditions simulating the stresses to which animals would be exposed in space flight. The concomitants that we have sought are those involving changes in brain function in alerting and arousal, sleep and unconsciousness, coma, epilepsy as it may arise from radiation, and the electric correlates of brain activity relating to the behavioral processes.

We have implanted recording electrodes deep within the brain (Figure 1), with the electrodes held in place by a mass of dental acrylic attached to the skull with stainless steel screws. Attachment of connecting leads is to two small plugs on top of the animal's head. Both cats and monkeys are normally active and alert after this procedure. One can even do it with man, although I would not suggest that it is the sort of thing one would wish to do to an astronaut. He might object. However, we are speaking now of the backup type of program from which we expect to get a certain type of data which would help to provide a base line for man's survival.

Experimental animals can be trained to do a vast variety of things. For instance, they can move freely in a T-maze training box with a cable attachment to its head, and can make an approach to the bottom of the box looking for a food reward on one or the other side of a partition. In these circumstances one can obtain from different deep-brain regions a variety of records. For instance, some regions show a great deal of very fast electric activity at about 40 waves per second, and this we have come to know is associated with arousal of the animal to the need for food or the opportunity for sexual reward and aggressive behavior. And quite a different type of record comes from certain other regions. Where trains of slow waves occur during the performance of a discriminative task, the animal is looking for food in a situation requiring some form of visual discrimination. These types of brain-wave records are, in fact, as specific as thumb prints in terms of the type of rhythm that comes from a particular region.

These different patterns of wave activity are recorded (Figure 2) in certain parts of the temporal lobe of the brain, which appear to

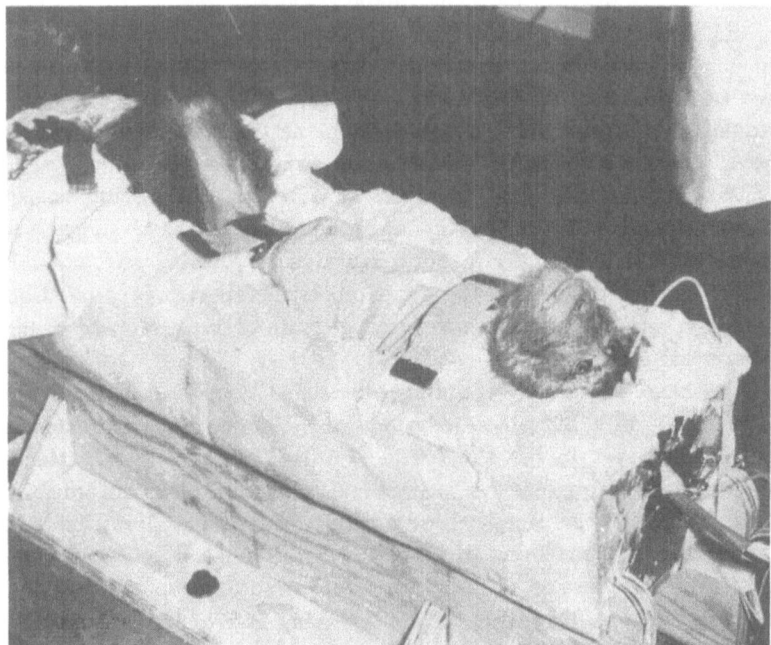

Figure 1.

relate importantly to the emotional state of the animal. For example, interference in these areas, by damaging small quantities of brain tissue, can lead to hypersexuality. Such small lesions will cause a cat to attempt sexual intercourse with a dog, or with a monkey, or with a chicken, or as many as four cats may engage in the delicate occupation of what is called tandem copulation. I am not at all sure whether there will be intercourse in space. It might become an important study in some future program.

In another situation, we have tested delayed-response behavior. Here the animal awaits food concealed under one of two identical cans for five to ten seconds before the animal can approach the concealed food. Characteristic rhythms appear both during the waiting period and during the subsequent approach.

Finally, we come to the question of what changes occur in the brain-wave records of cats and monkeys exposed to centrifugal accelerations of 8 to 10 g, and to shaking stresses. During accelerations in axes transversely to the long axis of the body, very regular slow rhythms appear in some parts of the temporal lobe at increasing

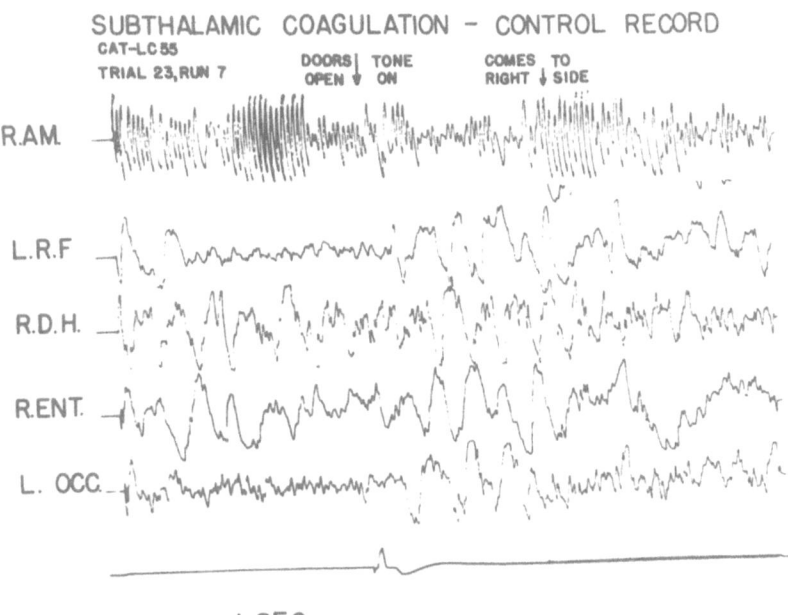

Figure 2.

transverse accelerations up to 8 g. Once the acceleration is sustained, however, the alerting rhythm disappears. As the deceleration commences, the rhythm comes back again, and it disappears as the centrifuge becomes stationary. If, however, one centrifuges the animal longitudinally to a blackout, among the last things to disappear as the animal loses conciousness are those little 40-persecond burstings that we spoke of earlier that appear in this part of the brain during extreme arousal for food and in profound alarm. This is the last thing to go from the normal record as it flattens out, and at this time some quite abnormal discharges appear, due to the loss of blood supply to the brain. This abnormal activity is of a type which we know to be an epileptic discharge. What would this do to a man's performance? Now in the first place he doesn't drop to the floor of the capsule and have a fit, in the classic sense. He would, however, be inaccessible if spoken to, and he would be quite incapable of exercising judgment. And this is, in fact, the most dangerous aspect of this type of problem, that this very severe acceleration could produce a loss of capacity to exercise a decision-making function.

When one puts a monkey on a shaker table, it is interesting that during shaking between 5 and 30 cps at relatively low *g* levels, between 0.5 and 3 *g*, the very powerful barrage of impulses set up from the muscles and joints appear to actually drive the brain rhythms in a very abnormal way (Figure 3). This abnormal driving we know to be associated with a loss of contact with the environment; to the point where it would probably be difficult for the man or animal to exercise a decision and make a correct judgment.

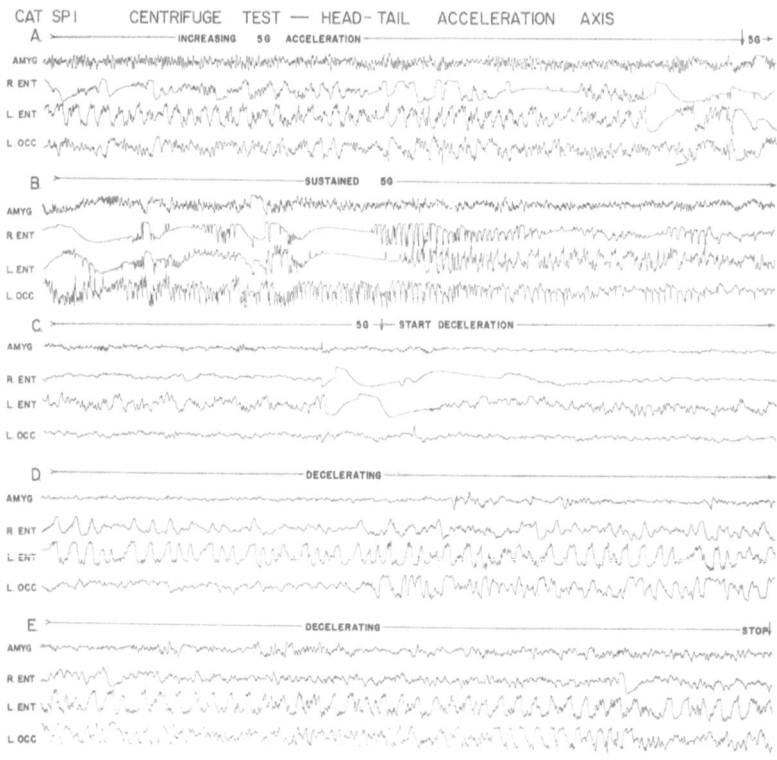

Figure 3.

In conclusion, these are a few of the changes that one might look for in a series of animal programs designed to test the feasibility of the survival of man under conditions of prolonged space flight. We might imagine that under conditions of prolonged weightlessness there would be changes in patterns of brain rhythms, reflecting the changes from wakefulness to sleep, and vice versa.

Such changes can be very easily discerned from this type of record. There is, finally, the question of the study and quantification of interpersonal relations of individuals in a group situation, particularly in the confines and constraints of the capsule environment, where defects in communication and rivalries between a group of closely knit individuals may provoke aggressive situations. In this field, also, brain-wave recording can provide a valuable monitor of potentially agressive behavior in animal studies performed as a backup to any manned space program.

Dr. Graybiel:[1] It is generally agreed that less is now known concerning the effects of "weightlessness" than in the case of any other environmental stress which man will experience in prolonged orbital flight. It is feasible to avoid weightlessness by causing the vehicle to spin, but this introduces two new problems. The first is concerned with how much centrifugal force should be produced and the second with "canal sickness." Because of the limitations in simulating weightlessness (and "subgravity" states), a satisfactory solution to this problem must await experimentation in space. It is possible, however, to study the problem of canal sickness by simulating the angular velocities which would be used in generating an inertial force field. The more completely this problem is investigated, the more precisely questions can be answered concerning performance aloft in a constantly rotating environment. The chief purpose of this report is to describe briefly the symptomatology of canal sickness, based on the results of experiments in which human subjects were exposed to constant velocities of 1.0 to 10.0 rpm in a small, completely enclosed room [1-4]. By way of introduction, I would like to review briefly the concept of "the gravitational-inertial force environment," and mention the problem of weightlessness.

The Gravitational-Inertial Force Environment

Man's gravitational-inertial force environment has its genesis in gravity, due to a central field factor and the accelerations man experiences as a result of change in velocity or direction of motion (Figure 4).

---

[1]The opinions and conclusions expressed by Dr. Graybiel are his own and do not necessarily reflect the views or endorsement of the Navy Department.

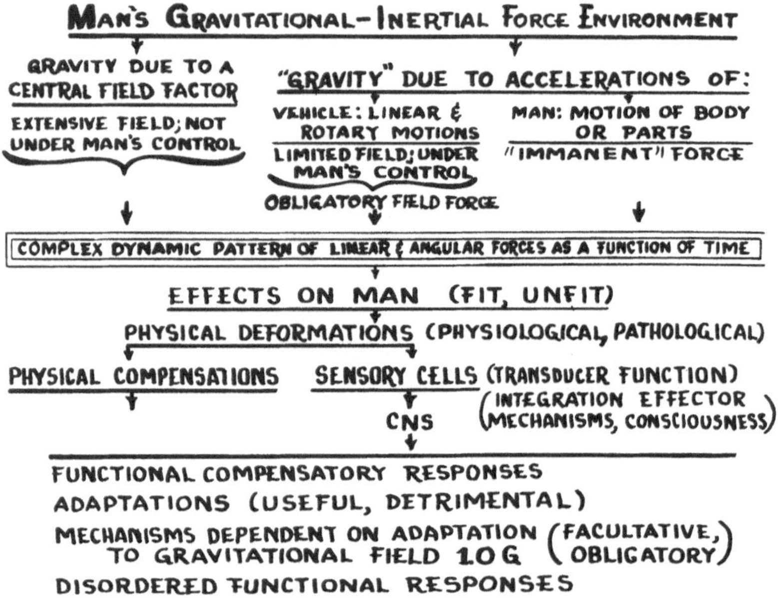

MAN'S GRAVITATIONAL-INERTIAL FORCE ENVIRONMENT

GRAVITY DUE TO A CENTRAL FIELD FACTOR	"GRAVITY" DUE TO ACCELERATIONS OF:	
EXTENSIVE FIELD; NOT UNDER MAN'S CONTROL	VEHICLE: LINEAR & ROTARY MOTIONS LIMITED FIELD; UNDER MAN'S CONTROL	MAN: MOTION OF BODY OR PARTS "IMMANENT" FORCE

OBLIGATORY FIELD FORCE

COMPLEX DYNAMIC PATTERN OF LINEAR & ANGULAR FORCES AS A FUNCTION OF TIME

EFFECTS ON MAN (FIT, UNFIT)

PHYSICAL DEFORMATIONS (PHYSIOLOGICAL, PATHOLOGICAL)

PHYSICAL COMPENSATIONS   SENSORY CELLS (TRANSDUCER FUNCTION)

CNS (INTEGRATION EFFECTOR MECHANISMS, CONSCIOUSNESS)

FUNCTIONAL COMPENSATORY RESPONSES
ADAPTATIONS (USEFUL, DETRIMENTAL)
MECHANISMS DEPENDENT ON ADAPTATION (FACULTATIVE, OBLIGATORY) TO GRAVITATIONAL FIELD 1.0 G
DISORDERED FUNCTIONAL RESPONSES

Figure 4.

Man is always exposed to a scalar gravitational potential[2] and, while his mass is countersupported, a gravitational force. Under ordinary living conditions on earth, it may be regarded as a constant, and as the only force of sufficient magnitude to affect total body weight significantly. It is the force to which man has become adapted throughout his evolutionary development, and to which he is accustomed through experience. The direction and sense of the vector representing gravitational force (g) indicates the upright, and the addition of mutually perpendicular lines forms a spatial frame of reference. Change in position of the body with respect to gravity introduces dynamic effects similar to those where the direction of gravity has changed with reference to man. The means by which man orients himself to the direction and force of gravity represent some of the most fundamental and important responses in his adaptation to life on earth.

The inertial forces generated by the active motions of the body or parts of the body may be regarded as "imminent" forces, inas-

---

[2]Theoretically, there are exceptions [5].

much as they do not contribute to the external force field. These forces are of small magnitude and derive their significance partly because they are associated with motions which change the position of the body with regard to the other components in the force environment, and partly because these forces are sufficient to stimulate sensory receptors which, through nervous mechanisms, aid in locomotion and in maintaining bodily equilibrium.

The introduction of artificial means of transportation and, in particular, aerospace flight has added the third dimension to man's problem of orientation, and has created new patterns in the force environments to which he is exposed. The magnitude of these linear forces may be so great that the force of gravity is small by comparison, and the angular accelerations may be far different in pattern from those ordinarily experienced. Combined, they constitute a complex, dynamic pattern which varies as a function of time. Although the equivalence of gravitational and inertial mass is the unifying principle underlying the force-environment concept, this simplicity gives way to great complexity when account is taken of the structural and functional characteristics of the body.

The effects of these forces have their origin in physical deformations (strains) and the ensuing physical compensations. Although the laws governing these physical changes are known, their complete pattern is difficult to determine because the body is not uniform, and because a state of mechanical equilibrium in all parts of the body is never reached. Moreover, these physical events initiate functional changes which may or may not be compensatory in the sense that they act to minimize or abolish the stresses. The best example is seen in the deformation of a sensory cell, which, acting as a transducer, gives rise to a nerve impulse. The impulses eventually reach many portions of the central nervous system, initialing effector responses which may or may not be compensatory in character.

The Problem of Weightlessness

If gravity due to a central field factor is counterbalanced by the accelerative force of the vehicle, the only forces to which man will be exposed will be those resulting from movements within the vehicle (Figure 5). Linear accelerations affecting gravireceptors will be sporadic and usually inadequate for deep sensibility — and probably always inadequate to stimulate the otolith organs. On the other hand, angular accelerations involving the head will stimulate the receptor cells in the semicircular canals. These peripheral organs will continue to function normally so long as their functional integrity is maintained. Any peculiarities in response to stimulation may be attributable to alterations in the usual "balance" of sensory inputs; however, this should be short-lived.

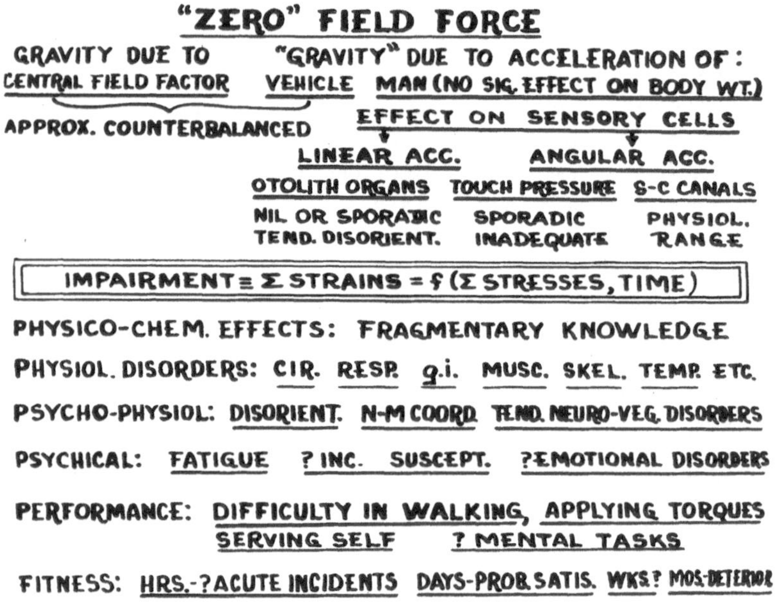

"ZERO" FIELD FORCE

GRAVITY DUE TO CENTRAL FIELD FACTOR	"GRAVITY" DUE TO ACCELERATION OF : VEHICLE MAN (NO SIG. EFFECT ON BODY WT.)	
APPROX. COUNTERBALANCED	EFFECT ON SENSORY CELLS	
	LINEAR ACC.	ANGULAR ACC.
	OTOLITH ORGANS TOUCH PRESSURE	S-C CANALS
	NIL OR SPORADIC SPORADIC TEND. DISORIENT. INADEQUATE	PHYSIOL. RANGE

IMPAIRMENT ≡ Σ STRAINS = f (Σ STRESSES, TIME)

PHYSICO-CHEM. EFFECTS: FRAGMENTARY KNOWLEDGE

PHYSIOL. DISORDERS: CIR. RESP. $q.i.$ MUSC. SKEL. TEMP. ETC.

PSYCHO-PHYSIOL.: DISORIENT. N-M COORD. TEND. NEURO-VEG. DISORDERS

PSYCHICAL: FATIGUE   ? INC. SUSCEPT.   ? EMOTIONAL DISORDERS

PERFORMANCE: DIFFICULTY IN WALKING, APPLYING TORQUES
SERVING SELF   ? MENTAL TASKS

FITNESS: HRS.-? ACUTE INCIDENTS DAYS-PROB. SATIS. WKS? MOS-DETERIOR

Figure 5.

Many physiological mechanisms either will not function or will not function normally under zero gravity conditions, but it is not known whether these losses or disturbances in function are of critical importance. In all likelihood, man will get along well for a period of hours and will tolerate exposure for days. However, his fitness would almost surely deteriorate over a period of months. The study of the effects of prolonged exposure to weightlessness can only be satisfactorily carried out in a space laboratory, and the same is true for subgravity states. However, one stress, namely, prolonged exposure to a constantly rotating environment, can be simulated in the laboratory.

Observations in the Slow Rotation Room (SRR)

If generation of an artificial field force aloft is desired, the angular velocity will probably fall in the range of 1.0 rpm where symptoms are almost negligible to 10.0 rpm where they are severe. The values in Table 1 indicate the radius of rotation necessary to produce 0.1 to 1.0 *g* at angular velocities between 1.0 and 10.0 rpm, the velocity range covered in our experiments. At very short radii the complicating effects of the exposure of the body to variations in centripetal force must be considered.

TABLE 1. RADIUS OF ROTATION NECESSARY TO PRODUCE
0.1 TO 1.0 g AT ANGULAR VELOCITIES BETWEEN 1.0 AND 10.0 rpm

rpm	Centripetal Force (g Units)									
	0.1	0.2	0.3	0.4	0.5	0.6	0.7	0.8	0.9	1.0
1	293	587	880	1174	1467	1760	2054	2347	2640	2934
2	73	147	220	293	367	440	513	587	660	733
3	33	65	98	130	163	196	228	261	293	326
4	18	37	55	73	92	110	128	147	165	183
5	12	23	35	47	59	70	82	94	106	117
6	8	16	24	33	41	49	57	65	73	81
7	6	12	18	24	30	36	42	48	54	60
8	5	9	14	18	23	28	32	37	41	46
9	4	7	11	14	18	22	25	29	33	36
10	3	6	9	12	15	18	21	23	26	29

The experiments were carried out in a closed room, about 14 ft in diameter and 7 ft high, constructed around the center post of the Pensacola Human Centrifuge (Figure 6). The room contained experimental equipment and living accommodations for as many as five persons. It was possible to control the rate of rotation within 2.5 per cent of any desired velocity between 1.0 and 10.0 rpm. Slip rings provided the means for transmitting electrical power or signals between the rotating room and the outside.

At constant angular velocity the subject was not exposed to angular acceleration so long as he was stationary with respect to the room. The centripetal force varied with the angular velocity and distance from the center of rotation; the maximum, at a radius of 6.5 ft at 10.0 rpm, was 0.22 *g*. If the subject moved (rotated) his head out of the plane of rotation of the room, the semicircular canals

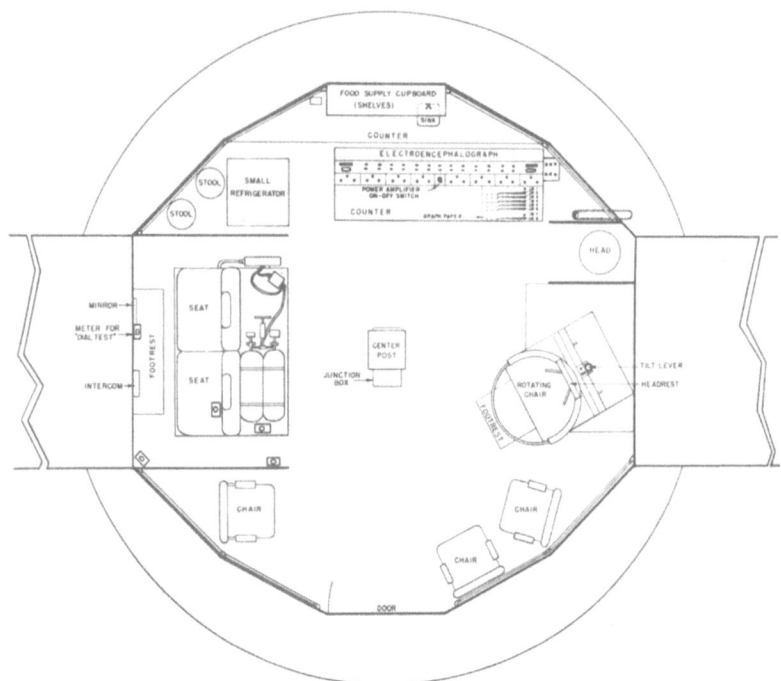

Figure 6.

were stimulated by virtue of a coriolis acceleration [6-9]. Guedry [9] has pointed out that the coriolis stimulus to the cupola is due to mechanical couples resulting from different magnitudes of coriolis acceleration at points in the canal which are not equidistant from the center of rotation.

This coriolis stimulus, which was a function of the angular velocity of the room and the rotation of the head out of the horizontal plane, was largely under the control of the experimenter. The angular velocity of the room could be controlled quite precisely, but there were limitations in controlling the head movements. One means of experimenter-paced rotations was the so-called "dial test" [1]. The subject was required to set the needle indicator of five dials so placed that different head movements were required to see the face of the dial and move the adjusting knob. A tape recording was used to pace the subject. Head movements associated with various bodily activities were partly under the control (restriction) of the experimenter.

It is necessary to mention that the periodicity, as well as the strength of the stimulus, was an important factor. Sinusoidal or other rhythmical motions of the head were better tolerated than aperiodic or irregular movements. The significance of this phenomenon in terms of neural mechanisms is under investigation.

The coriolis accelerations were not the only factor determining the subject's responses. Environmental factors included room temperature and humidity, odors, and the visual field. Physical fitness was a factor, and the advent of an upper respiratory infection increased the susceptibility to symptoms in one subject. Psychological factors were very important, but elusive. Anxiety based on past experiences of motion sickness, or apprehension based on reports of other subjects, was almost surely significant in some instances. Also, it was found that speeding up the pace at which the dials were set often decreased the manifestations, despite the increase in stimulation to the semicircular canals — presumably the result of increased attention to the activity. Persons free to carry out activities, but complaining of symptoms, might improve when given specific tasks to carry out involving greater stimulation of the semicircular canals.

That the symptomatology had its genesis in stimulation of the semicircular canals was demonstrated by the fact that one subject [1] who had lost all function of the sensory organs of the inner ear experienced none of the illusions or unpleasant symptoms experienced by subjects with normal sensitivity. Moreover, two subjects with moderately decreased sensitivity of the canals experienced no unpleasant symptoms, but had difficulty in walking [10]. Preliminary evidence suggests that the otolith organs do not play an etiological role. Undoubtedly, the unusual patterns of afferent impulses originating in the semicircular canals disturb central nervous system mechanisms and result in a wide variety of responses (Figure 7). A more complete listing of the subjective and objective response is shown in Figure 8.

Canal Sickness

The term canal sickness has been proposed to designate this etiologic type of motion sickness [11]. It may be defined as a syndrome which is experienced by all persons with normal function of the semicircular canals exposed to unusual patterns of angular accelerations of sufficient magnitude and duration. Predisposing

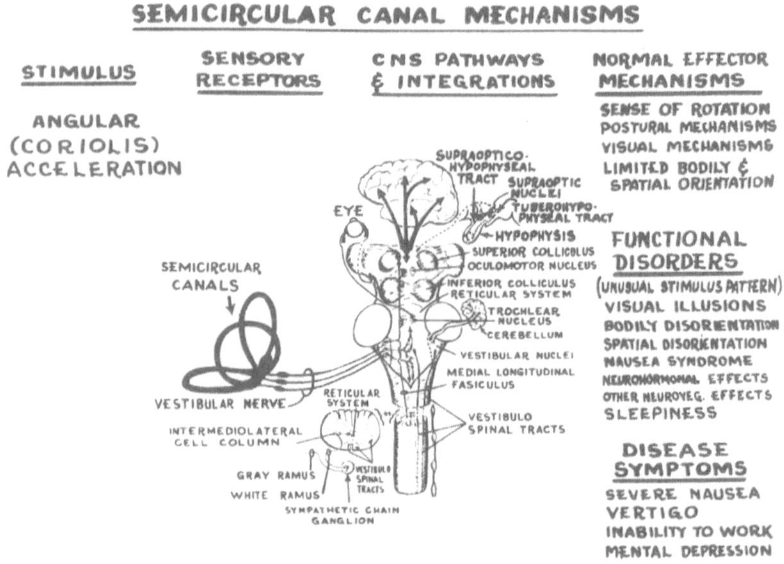

SEMICIRCULAR CANAL MECHANISMS

Figure 7.

factors can be categorized under basic individual susceptibility, state of fitness, pattern of other sensory inputs, mental alertness, and unfavorable environmental factors. Cardinal symptoms are visual and postural illusion, sweating, nausea and vomiting, somnolence, apathy, and difficulty in walking. Canal sickness is a functional disturbance, and symptoms may quickly disappear after cessation of the causative stimulus; sometimes they persist for many hours. Adaptation occurs in the course of hours or days in all subjects, although it may not be complete. Habituation also occurs in the sense that repeated exposures tend to lower the susceptibility to symptoms.

For similar activities (head movements) the severity of the symptoms closely paralleled the angular velocity of the room. At 1.0 rpm even highly susceptible subjects experienced few symptoms, and these disappeared in the course of a few hours [4]. From 1.71 to 5.44 rpm, there was a progressive increase in severity of symptoms, but even the most susceptible subjects manifested adaptation to the constant rotation within a period of two days. Above 5.44 rpm, there was a sharp increase in severity of symptoms, and even nonsusceptible normal subjects were unable to adapt completely within a

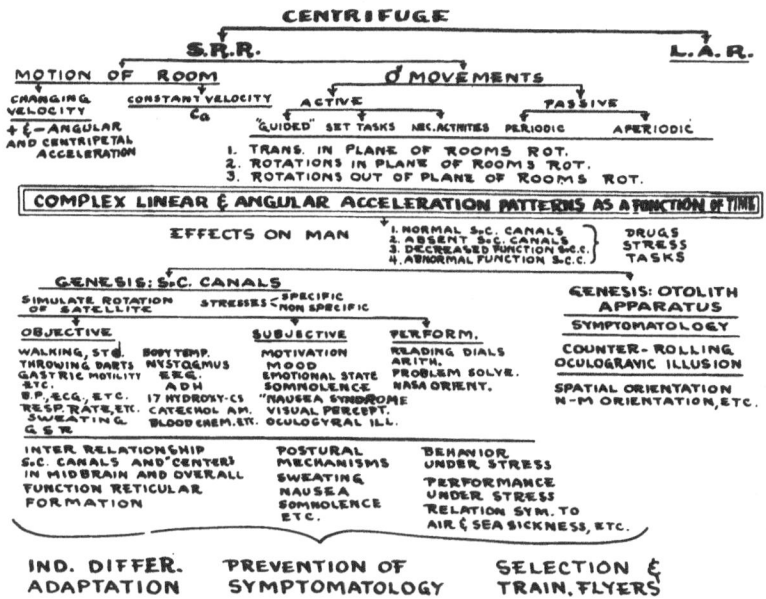

Figure 8.

period of two days. Subjects with moderately depressed function of the canals were relatively little affected even at 10.0 rpm.

The fact that persons who have lost the function of the canals do not experience any unpleasant symptoms forms a point of departure in an attempt to prevent canal sickness. Indeed, even a moderate decrease in function of the canals appears to be sufficient. Moreover, this small decrease does not appear to handicap the person in carrying out any ordinary activities. The reason for this lies in the fact that the positive contribution of the semicircular canals to man's powers of orientation and maintenance of equilibrium is not important except in the absence of vision. This stands in sharp contrast to the very great potential contribution which the canals may make in terms of disturbed functions. These disturbances may arise either from pathological causes or as a result of bizarre patterns of stimulation.

In this connection, the findings in cases of Meniere's disease treated with streptomycin are of interest [12, 13]. It has been found that administration of relatively large amounts of the drug abolishes the function of the canals, as determined by caloric testing, without any deleterious effect on hearing. The only troublesome side-effect was ataxia for a period of weeks. Although these results cannot be applied directly to persons with normal sensory organs of the inner ear, they do form a starting point for experimental studies on animals and ultimately man. Hopeful possibilities include the likelihood that only a moderate decrease in threshold of sensitivity of the canals would suffice, that the temporary associated ataxia would not be handicapping, and that related drugs might have even a greater elective affinity for the receptors in the cupola than streptomycin sulphate.

Even in the absence of any direct approach involving the peripheral end organs, much can be done to minimize the symptoms of canal sickness by means of selection, habituation, restriction of head movements, minimizing predisposing factors, and the use of drugs.

Conclusion

In conclusion, it is worth emphasizing that the procedure involving the use of the slow rotation room offers opportunities to study not only the role of the semicircular canals under certain unusual patterns of stimulation encountered in aerospace flight but, also, the many neural mechanisms involving particularly specific centers in the brain stem and some of the functions of the reticular formation. It may be useful in the selection, indoctrination, and adaptation of flyers. The fact that even a moderate increase in threshold of sensitivity of the receptors in the semicircular canals prevents the appearance of symptoms, or greatly reduces their severity, suggests a hopeful lead in terms of selection of "unsusceptible" personnel and preventive treatment. The control over the stimulus situation and the resulting widespread symptomatology hold forth good possibilities for studying some of the truly remarkable neural mechanisms involved in the exhibition of symptoms and the phenomenon of adaptation.

REFERENCES

1. A. Graybiel, B. Clark, and J. J. Zarriello, "Observations on Human Subjects Living in a 'Slow Rotation' Room for Periods of Two Days," *Arch. Neurol.*, 3:55-73 (1960).

2. B. Clark and A. Graybiel, "Human Performance During Adaptation to Stress in the Pensacola Slow Rotation Room," *Aerospace Med.*, **32**:93-106 (1961).

3. A. Graybiel, F. E. Guedry, W. H. Johnson, and R. Kennedy, "Adaptation to Bizarre Stimulation of the Semicircular Canals as Indicated by the Oculogyral Illusion," *Aerospace Med.*, in press.

4. R. S. Kennedy and A. Graybiel, "Symptomatology During Prolonged Exposure in the Pensacola Slow Rotation Room at a Velocity of One Revolution per Minute," in preparation.

5. H. Strughold and O. L. Ritter, "The Gravitational Environment in Space," in O. O. Benson, Jr., and H. Strughold (eds.), *Physics and Medicine of the Atmosphere and Space*, New York, Wiley, 1960.

6. J. E. Purkinje, "Beitrage zur naeheren Kenntnis des Schwindels aus heautognostischen," *Daten. Med. Jb.* (Vienna), **6**:79-125 (1820).

7. E. Meda, "A Research on the Threshold for the Coriolis and Purkinje Phenomena of Excitation of the Semicircular Canals," translated by E. R. Hope, *Arch. di Fisiol.*, **52**:116-134 (1952); translation T 17 1, Defence Scientific Information Service, DRB Canada (1954).

8. G. Schubert, "Die physiolischen Auswirkungen der Coriolis-beschleunigungen bei Fluggeugsteurung," *Z. Hals-Nas.-u. Orenheilk.*, **30**:595-604 (1932).

9. F. E. Guedry and E. K. Montague, "Relationship Between Magnitudes of Vestibular Reaction and Effective Coriolis Couples in the Semicircular Canal System," Report No. 456, Fort Knox, Ky., U.S. Army Medical Research Laboratory, 1960.

10. A. Graybiel and J. C. Meek, "The Increase in Threshold of Response of the Semicircular Canals Sufficient to Prevent Canal Sickness in Man," in preparation.

11. A. Graybiel, "Orientation in Space with Particular Reference to Vestibular Functions," paper presented at the 1958 International Symposium on Submarine and Space Medicine (New London, Conn.).

12. P. Northington, "Syndrome of Bilateral Vestibular Paralysis and Its Occurrence from Streptomycin Therapy," *Arch. Otolaryng.*, **52**:380-396 (1950).

13. H. F. Schucknecht, "Ablation Therapy in the Management of Meniere's Disease," *Acta Otolaryng.* (Stockholm), Suppl. 132, 1957.

Dr. Singer: There is considerable disagreement over the role of man in space. On the one hand, you have the enthusiasts – the "space cadets." They would have man go out immediately setting up colonies on other planets and moving on to other stellar systems using the method of Noah's ark. Physical scientists, on the other hand, have been much more cautious. Too much so, I think. The antithesis which they take is the point of view that there is nothing a man could do which cannot be done better by a little black box – you know, a little device that performs certain functions. It's done with electronics; you don't need to know what's inside of it. It's a black box, it does what you want it to do. The scientists have been perhaps not as vocal as the space cadets, still they have been sounding off too. Now I would like to get into this discussion by moving in toward the center. I think a moderate approach is necessary here and not the extreme approach advocated by both groups. I think even space enthusiasts are becoming aware of the fact that the role of man in space has to be defined somehow. You can't just say, "Well, let's go out in space." You do have to give some valid reason, make some attempt to justify it, if only in order to get funds. Physical scientists, I think, are well aware of the fact that man is going to be traveling in space anyhow, and, therefore, we might as well think about his proper functions. I think, in fact, it makes very little practical sense for anyone to argue that man should not go into space. The machinery has been set into motion. There is nothing you can do to stop it – which reminds me of a story about Mark Twain. He was having a big discussion with an old riverboat pilot on the Mississippi, and they were just changing over from sail to steam in those days. This pilot was fussing and arguing that he didn't like these newfangled inventions. He thought sail was very much better, and he wanted to stick with it. Mark Twain after a while got fed up and looked at him and said, "You know, when it's steamboat time, you steam." Well, it's steamboat time, so we have got to think up some good valid functions for man in space. But there is one thing that is fairly certain, whatever we decide right now that man is going to do in space, the chances are he will probably find other things which are much more important, and which we can't predict yet. In view of the fact that no man has as yet traveled in space, it seems almost futile to predict his functions or even to discuss his proper role, *but it is not quite futile.* We have to do this, and it is more than an exercise because the *justification for a*

manned space vehicle has to be taken into account when one de-
signs a functional vehicle from a physical point of view. Therefore,
the discussion of man's role in space does have considerable practical importance for the design of the immediate vehicles.

Now, if I were to stress the major advantage of man over the
black box, it certainly would be man's *flexibility.* A man can perform many more functions than any black box which has as yet been
conceived. This is quite natural, since the human brain is much
more capable of many programming situations – speaking in electronic terms – than any known computer. This is not to say that a
black box may not do a particular function better than a man, but
man will always excel in his versatility to perform a variety of
functions and in his flexibility to improvise and to adapt to the unexpected. I think this is the important point. Whenever you run into
a situation which involves the unexpected you cannot, by definition,
design a black box to deal with it, because you don't know what to
expect. I think next in importance is man's selective perception.
This ability to filter data, to look at all the data, but to discriminate
and only select the important from the unimportant, is vital. And
finally, his ability to reason and exercise judgment after he has
absorbed all of the data is also crucial.

A remark was made by Crossfield that "man is the only nonlinear,
150-pound servomechanical system which can be mass-produced by
unskilled labor." This is a tremendous advantage when you are
trying to do a complicated job. It is much more expensive to design
a computer to do the job than to use a man. Of course, the disadvantages of man are well known; for example, his general fragility
and his dependence on a very complex life-support system, his
inability to do a monotonous task very well for a long period of
time, and his inability to perceive many things which instruments
can detect. For example, a man cannot directly detect cosmic rays,
but a Geiger counter can. Nevertheless, we may expect that man and
these black-box instruments will complement each other, and
each will find its proper role. Whenever I get involved in a discussion with some of my colleagues who argue violently against
man in space, I always tell them, "Well, if the good Lord had wanted
to, He could have made black boxes instead of men." And if they
still want to argue, I always say, "Well, I don't argue with atheists."

When it comes to making scientific explorations and observations, I think man is at a disadvantage with respect to instruments.

But there is one field of scientific observation for which instruments are not appropriate. These are observations on man himself as a biological system removed from his planet. We have never studied man in the absence of the earth. Man has always been in a terrestrial laboratory, subject to the gravitational pull of the earth, subject to the stimuli which go with living on the earth. Now, of course, you have heard the discussion here by the preceding speakers who have mentioned laboratory situations in which space environment is simulated. Some people have conceived systems whereby a number of these environmental factors could be produced right here on the earth without sending a man up into space, or before sending a man up into space, and this is certainly most valuable. But it is almost certain that we cannot really simulate every one of these factors. Maybe we will succeed in simulating the important ones. And certainly what we learn about man in space, as man is a biological system, will be useful to us, because we are men and we can then apply this information to ourselves and for our benefit.

Now one of the items, of course, which intrigues us most is the general subject of life on other planets. As a physical scientist I cannot discuss it biologically, but I am convinced this question will finally be solved in our time by experimental means, that is, by landing instruments on the planets and then by landing man himself. Life, as you know, has its origin in very elementary forms, starting with simple molecules and ions and proceeding then to more complicated molecular chains. As you know, the molecular chains for earth-based life systems have a certain helix arrangement. They may also be based on different elements. In any case, if these elementary organic units appeared elsewhere, they would exist under conditions different from terrestrial ones. There is nowhere, in our solar system at least, any comparable condition to the one existing here on the earth. These organic combinations would have evolved in millions or billions of years, and would have adapted themselves to conditions existing on other planets. For example, life may have made its appearance on Mars. We know that some form of life does exist on Mars, perhaps a very low order of vegetation. This may have appeared several billion years ago and evolved in some form capable of existing under Martian conditions. These conditions would be absolutely deadly to any terrestrial organism which arrived without necessary protection. This is why we have to protect man; even after he has landed on the planet he still has to be supplied with an earth environment. Now if life on Mars, however, has developed to the point of formation of thinking beings who have reached a certain level of scientific knowledge, these beings might very well believe that life does not exist on earth, just because conditions here are different. They might think, for example, the atmospheric pressure on earth is much too great for any living being to survive. Our views of life on other planets are colored by the same types of prejudices. We are very close to a decisive turn here as regards our conception of life, its origin and its development — which brings this discussion back to the role of man in space.

I think one of the really exciting prospects for exploration is, of course, this discovery of a form of life which has developed apart from development on earth. It may be quite difficult to recognize such a form of life. One of the chief functions of man's exploration on a planet might be the use of the peculiar facilities of the human brain to recognize life forms which could not be detected by an instrument. This function of man as an explorer, not only of the geography and environment of planets, but of the existence of life forms, may constitute one of the really important reasons for manned space flight. I am trying here to think up various good arguments — reasons as to why manned exploration of space is really desirable. It is difficult; you cannot readily find all the situations in which man can outperform the black box, but I think this is one in which a black-box design certainly would be very, very difficult, very uncertain. If we wanted to sample or explore the geography of a planet — let's say the moon — we might land equipment there which would sample the area in which it lands, and send back to the earth a little piece of the lunar surface. But we might miss a very important fact which a man would obviously see if he were there, but which the equipment doesn't see because it only samples the area in its immediate vicinity. Another application is that the data-transmission system for complicated exploration may be so involved as to be quite expensive or even not feasible. For example, if you wanted to recognize life, an instrument would have to perform many, many important and delicate tests — all sorts of tests — and this would constitute a great deal of data — a great deal of information — which would then be transmitted to the earth. On earth a man would sit and look over all the data and then decide, yes, life does exist, or no, it does not exist. How simple to have a man up there looking at the situation and simply transmitting back yes or no. It saves a lot in the data-transmission system.

As far as physical observations are concerned (I am talking now about observations of radiation, magnetic fields, and the like), man will always play second fiddle to instruments. I don't see any way of really beating a good magnetometer. We are just not equipped to detect magnetic fields. But we can be sure of one thing: that sooner or later the instrumentation systems which will be built to perform certain space functions will become so complicated and expensive that we will need man, but in a very unglorious role as a sort of maintenance man, a repair man. For example, we may have a very complicated television and communications satellite, something in which a lot of money has been invested, or a complex astronomical observatory in space. If something goes wrong with the equipment, it will take somebody who will be an adjuster, or some kind of device which is cheaper to produce than the black box, and that will be a man. He will go up there and fix it. And it's a perfectly reasonable application for a man because he can make full use of his facilities, his judgment, his reason, and his ability to improvise. If the satellite is inoperative, it will have to be repaired. The equipment which is faulty may have to be replaced, lady space travelers in distress may have to be rescued, and so on.

Now there is one general field that I would like to touch on briefly, and this is the military implications. The military field is, of course, the one where man would assume his most important role. I think there is no doubt that reconnaissance, for example, has an important function. A man might be an ideal mechanism for providing the judgment, selection, and filtering which is necessary to make a reconnaissance system operate most effectively. I am not saying that he would replace the television system, but he may operate as a very valuable adjunct in addition to it. Quite a different type of reconnaissance involves the inspection of unknown satellites to establish their nationality, their identity, and their purpose. The satellites may have to be boarded, further inspected, captured, disarmed, and possibly destroyed. Again this is an important function for man. But interestingly enough, man himself is very vulnerable in space. Not only is he very fragile and completely dependent on carrying his environment with him, but also his molecules are long-chain types and, therefore, extremely sensitive to nuclear radiation both natural and artificial. Atomic explosions in space, whether they are naturally produced on the surface of the sun, or whether they are made by weapons, produce nuclear radiation of sizable intensity that will destroy a man very quickly. So whatever new military concepts may evolve for man's operation in space, they will have to take into account this tremendous fragility of human beings.

DISCUSSION

Mr. Mayo: The introductory statements indicate quite clearly that the slant of this particular panel is somewhat to the proponent side of center with respect to whether you put man in space or not. I get the impression from each of the discussions that there is a significant need. Even our physical scientist in the group has talked rather strongly on this point. One very significant fact is that the human brain is a part of every space system in the final analysis; no knowledge is available except by means of a human brain. Consequently, even if we are talking of the Geiger counter, or the device measuring energy completely outside of the spectrum of any of our direct sensory systems, we must somehow conform the data to a form acceptable to human senses so that information gets into the brain in order to become useful knowledge. For this reason, some place down the line the link must be made. Therefore, we must realize that here is an important clue toward the resolution of the problem of where a man belongs in a given system. Let us first have any comments from the panel, and then consider questions from the audience.

Dr. Orlansky: There was such a remarkable degree of uniformity among the members of the panel that I should like to shift my position 180 degrees. The original question was whether a space laboratory would improve our knowledge of human performance in space. As a matter of scientific interest, each of the speakers answered yes. Let me change my answer to yes, but....

Suppose we ask whether it is a good idea to put man in space. This may be approached as a matter of good engineering practice. Fundamentally you specify what you want to accomplish and then try to identify the functions and components that are required to accomplish this purpose. Man may be regarded as an available component, with advantages and disadvantages that must be related to the particular purpose. Certain required functions in a space vehicle, such as attitude control, can be performed more effectively by an internal gyro system than by a man. As long as man is present in a space vehicle, he requires a life-support system. Obviously, this support equipment is not justified unless man provides the only way to accomplish an important task.

On the other hand, certain tasks can only be done by a man. Then it is necessary to bring man along despite the extra equipment which he requires to stay alive. His unique abilities include flexibility, decision-making, the ability to repair equipment, and so on; we do not know how to perform these functions in any other way.

It is not always desirable to have man in a space system. The balance sheet must be carefully drawn for and against him in each particular case.

Having said this, I would like to introduce a further note of caution. Though man will certainly go into space, we still have an obligation to ask where we can best use our resources. We face many challenges and not all of them are in space. I would like to list just a few, such as cancer, or automobile accidents (40,000 people die this way each year), or mental illness. Further, I personally believe that the problem of arms control is more important for our survival than the question of man in space. In turning 180 degrees, then, I hope we direct our energies in an important direction, and this is not *necessarily* space exploration.

Professor Adey: This question of measuring the price of the need to put man into space is in itself a little presumptuous, because we don't even know yet whether man can go into space. In other words, we do not have the base-line data on which to make the decision as to the most appropriate means of protection to provide for man in all the hazards that he will be faced with in this type of experiment. Not at least until that type of data is available, either from direct observation on man himself or extrapolation from much more sophisticated animal experiments than measuring the rectal temperature and taking the electrocardiom, *ad infinitum*, can we discuss further issues. These are things that are not contributory to an understanding of the feasibility of putting man himself into space. The second point, I am sure, is one that has had a great interest for all of us, namely, is it reasonable to spend so much money on trying to put man into space when there are other areas of biological experiment that need very urgent investigation? I think the answer is that it should be more widely known that the support to fundamental research by the armed services in areas which are directly and indirectly connected with the space program has provided the equipment which is used very, very widely, indeed, in many areas of fundamental biomedical research. I need go no further than my own laboratory, where, after two years of clamoring at the doors of the National Institutes of Health for support to provide minimal data-reducing equipment, tape recorders, and so on, of the type necessary for our sort of research, we have received this equipment not from the National Institutes of Health but from the United States Air Force.

Mr. Mayo: I might add that we see a large number of simpler ecological systems intervening between those presently required and the application of a closed ecological system. I can say though that a considerable amount of work is under way. This work will undoubtedly continue, and I think that we will see valuable results, particularly as we begin considering such problems as lunar bases.

Question: First, to what extent are sources of information being shared about the feasibility and possibility of space travel?

Dr. Singer: I'm elected since I don't work for the government and I can speak freely. Political considerations are, of course, involved — and rightly so. They play a part in any activity which the government carries on and which is supported by the taxpayer, and it is one of the considerations, along with scientific ones. I think in every person the mixture of political versus scientific varies; in some the balance may be fifty-fifty, in another, it may be ninety-ten, one way or the other. I think it's very obvious that the reason our man in space program is getting such a big boost in terms of funds is because of the efforts the Russians are making in this direction. And it's appropriate — I think quite appropriate — that we should have the capability of putting a man into orbit if not simultaneously with the Russians, at least very soon thereafter. I don't think we are going to lose very much momentumwise, although prestige unfortunately enters too. But I don't regard the political consideration as the overriding one. I think it is really scientifically very important that we do put man in orbit to learn the things about him that the, various panel members have indicated. It just happens that, due to the political situation, a lot of funds, a lot of steam, a lot of energy, are available for this particular project.

Question: More specifically, I asked to what extent is information being shared to make it possible?

Dr. Singer: I don't think any information is being shared which makes it possible. I don't think it matters very much, frankly, because I don't think the Russians can tell us anything about putting a man in space, since they have not done it themselves. I think if they do put man in space and learn something which is very unusual, they will publish it and we may take advantage of it. However, we don't take advantage of everything they publish.

Mr. Mayo: Does anyone else want to answer the political question? Did you get an adequate answer?

Question: It was not a political question except in the way it was answered. I asked, more specifically, to what extent information is being shared.

Mr. Mayo: I presume that you are referring to the Russians in this case and the answer, I think, is correct — practically not at all, with the exception of the International Geophysical Year cooperation. They have indicated a very great reluctance to this sort of thing. In fact, I was talking to Dr. Sedov in August with respect to the simple sharing of information concerning means of sterilizing payloads, to protect the scientific knowledge in which gains to either side would be gains to both, and he answered by asking if our capitalistic companies would be willing to share their trade secrets. This indicates, apparently, a considerable lack of desire for this type of cooperation.

Question: [Transcription unintelligible]

Mr. Mayo: The question was, "Will men go to the moon in a state of suspended animation or will they be working parts of the crew?" Dr. Adey, would you like to answer first?

Professor Adey: From the neurochronological point of view — and, again, this has tremendous strategic significance viewed from the point of view of national defense and in terms of international relations — what can be done with modern tranquilizing drugs and things of the like? It is perfectly feasible to suspend activity of a person for a determined or predictable period by drugs which act specifically in well-known brain regions, and apparently have no permanent effects and certainly no hangover. I personally cannot imagine that too many potential astronauts would wish to submit to this type of treatment, and I'm told that those who are interested in this sort of thing would prefer not to be drugged, as it were. At least this is the newspaper account of their attitude.

The question of whether it would enhance their ultimate capacity to work elsewhere is probably the crucial one, and I don't think that it has any bearing on the question. I think that, if they survive the buffeting, and so on, the stresses that are involved in this temporary type of stress would not debilitate them in any way in their ultimate performance. On the positive side, there is the fact they would not be in a position to press any panic buttons. They would be consigned to their fate by those on earth or in near space – if they were being sent to the moon or somewhere similar.

Mr. Mayo: We ought to have a psychological view on the question.

Dr. Orlansky: My psychological view is purely a personal one. The question is, "How should man go to space?" I hope he will go willingly, provided there is a reason for him to go and that he can do a useful job. I see no reason to carry him along as baggage. On a long voyage, should a man use drugs to sleep in order to reduce the food supply, for example? A man might agree to do this so that he would perform well when he was really needed at a later time.

Mr. Mayo: Dr. Singer, I think, has a point to add to this particular area, and perhaps a counterview.

Dr. Singer: From the physical point of view, to rephrase or expand your question slightly, I would like to put it in terms of a man going to, say, Jupiter, where the travel time is several years from the earth. You might ask a very valid question: If it is his scientific purpose to explore the surface conditions of Jupiter after he lands, then why keep him awake for several years? Why not, in fact, tranquilize him, assuming you can do this, slowing down his metabolism so that he will not require much air, much food, and other conveniences, including reading matter and the like, during these many years when he is completely inactive and has no useful function to perform? There is a second point of view, too. As long as he is awake and acting normally, he is, of course, aging at the normal biological rate. It is a great sacrifice to throw away several years of your life just to spend several hours on Jupiter. It would be much better essentially to have him, if you can, in a state of suspended animation, so that he would not age as rapidly biologically, and therefore save several years of his life. I think it is a solution which would be a very fair one to an astronaut if he is going on long voyages. However, I am presupposing that my biologist friends can solve this problem.

Professor Adey: I would just like to ask Dr. Singer what his evidence is that you don't age when you are tranquilized. Do you have any pharmacological or other data to support this point?

Dr. Singer: No, I don't. This is why I can speak with complete confidence. Let me carry on in this nonserious manner. I am not inhibited as you are by your training in biology and physiology; I am speaking outside of my field, so I don't have to be as careful as I usually am. I will then just say that I presume that — and this is a serious question really addressed to you — if the metabolism is slowed down, let's say by being in ice water or whatever other means you use, would you not be turning over your cell tissue more slowly? Would you than not be aging more slowly at least by one definition of aging?

Professor Adey: Well, it's a very good question, and it gets into the general problem of geriatrics, which in fact suggests that this very slowing is at the core of the aging process as is the laying down of some of these queer interstitial tissues, such as collagen and allied substances. The deposition of this material is not slowed, as far as I know, by slowing the general body metabolism. In fact, one of the very overt aspects of modified body function that appears when, for instance, the thyroid gland is interfered with to the point where it is functioning inadequately is increased deposition of these tissues. There is an appearance of premature aging, and the life span of people so born or so reduced is actually less than that of normal people. Some say this is the result of intercurrent infections to which they are particularly prone. But this is not the total answer. It may be that a person's chances of survival, by reason of many, many complex aspects of endocrine function — which would include the whole galaxy of steroid functions from his adrenals, the pituitary mechanisms at the base of the brain, and so on — may ultimately be best if he is in a wakeful state rather than reduced to some state of coma over a very prolonged period. This aspect of tranquilizing to the point of complete nonperformance over a long period is something about which not a great deal is known. Certainly, my psychiatrist colleagues, with characteristic enthusiasm, will drug people with doses 100 times greater than those which pharmacologists advise, thereby exposing the liver to extreme distortions of function, as our friend here mentioned. But leaving aside such aberrations of therapy, the question is: "Can one sustain life over very long periods by this method and expect at the end to have an individual capable of good function?" I am not at all sure that we can at the moment.

Dr. Singer: One additional question, again as a physicist. Let's look at an organism as just a biochemical system; we know that rate processes in chemistry depend very strongly on temperature. It would seem to me that if all the rate processes were slowed down proportionately by lowering the temperature, then you would be aging at a lower rate. In other words, you are then just expanding the time scale. Is that not so?

Professor Adey: Well, I don't know. But I suspect that the situation is by no means as simple as this. One has certain irreversible processes which gradually catch up with the total organism. In the state of reduced metabolism the approach of some irreversible process, as, for instance, damaged kidney function, or modified cardiac activity, or simply disturbed gastrointestinal functions, may be most insidious. All these processes can be thrown into some irreversible aspect of function which is not compatible with the continued life of the individual. This applies even to the temperature-control processes themselves. For instance, if one lowers the body temperature below 89° Fahrenheit for more than about two hours, it is very difficult to restore normal function. It is difficult to know just what is the exponential of this curve. In other words, if one lowers it three degrees, or four degrees, how long can it be sustained before such factors as the loss of intracellular potassium, and so on, make it incompatible with the revival and survival of the individual?

Question: Getting back to the question of the evening: "Is there a need for a manned space laboratory?" Would we not gain a tremendous amount of knowledge of our particular galaxy, as well as other galaxies, by observations in the astronomical sciences? And would we not gain a tremendous amount of knowledge of geophysics of the earth and, by application, of meeting the abnormal environments of space? Would we not gain knowledge to conquer the arid and bad areas on earth so that they would be more habitable?

Mr. Mayo: Well, those do not sound like questions, but part of the conclusion, and I'm very happy you saved me the trouble of bringing them up. I think it was very well stated. I couldn't do it better. One more question.

Question: The little black boxes are made to answer questions, and I wonder what is the use of putting man in space if he can generate questions to be answered.

Mr. Mayo: I think that was also very well stated. The black box does not generate questions, it only answers those specific questions it has been programmed to answer. This is one of the key points.

Question: [Transcription unintelligible]

Dr. Graybiel: I am not quite sure I heard all the question. You mean if the floor had been at a slant so that the direction of the resultant force was head to foot, would that have made a difference? Actually, the centrifugal force on the person was extremely small and near the center of rotation; I mean, it was barely perceived. At 1 or 2 rpm you are hardly aware of rotation, and at 10 rpm, when you are nearer the outer periphery of the room, the maximum g or the maximum centrifugal force was a little over 0.21 g so that it was actually a very small component. And this was experienced, of course, by the man without the semicircular canals; it didn't bother him. So I think it is a relatively unimportant factor in those experiments.

Mr. Mayo: We are drawing near our quitting time here. Perhaps we should defer any remaining questions to individual panelists until after adjournment. I will try very briefly to bring out one or two points in summary as to what I gathered was the consensus of the panel. A need in space exploration is to improve the amount of knowledge gained per unit of effort expended. In the final analysis, this knowledge gain must utilize the brains of human beings. To the extent that we can utilize human astronauts, as well as remotely controlled space vehicles, in maximizing results, we have an important application for manned vehicles. Certain of the studies cannot be conducted on earth because of the inability truly to simulate space environment. Externally, man is very fragile, and the only environment in which we have reasonable proof of his capability of performance is that of the earth. To the extent that we can study man in the achievable environment of a space vehicle, we can provide ourselves with a firm base of knowledge concerning what we can do with man in further work. It would appear then that *there is great need for a manned space laboratory* for a number of specific reasons, largely centering around the problem of the acquisition of highly detailed knowledge, and of the study of man himself. In this latter case he may be both a subject and an observer.